Central and Eastern European Development Studies

Bernhard Müller
Maroš Finka
Gerd Lintz
Editors

Rise and Decline of Industry in Central and Eastern Europe

A Comparative Study of Cities and Regions in Eleven Countries

With 29 Tables and 12 Maps

Springer

Volume Editors

Professor
Dr. Dr. h.c. Bernhard Müller
Leibniz Institute of Ecological
and Regional Development
Weberplatz 1
01217 Dresden, Germany

Professor
Maroš Finka, PhD
Institute of Spatial Planning
and Urban Design, Faculty of Architecture
Slovak University of Technology
Nam. Slobody 19
81245 Bratislava, Slovakia

Dr. Gerd Lintz
Leibniz Institute of Ecological
and Regional Development
Weberplatz 1
01217 Dresden, Germany

ISSN 1614-032X
ISBN 978-3-642-07342-7 e-ISBN 978-3-540-26695-2

Springer is a part of Springer Science+Business Media GmbH
springeronline.com

Printed in Germany

Cover design: E. Kirchner, Heidelberg

Printed on acid-free paper 30/3141/as 5 4 3 2 1 0

Preface

The political and economic face of Europe has changed dramatically over the past fifteen years. Most notably, the division into a western and eastern bloc has been overcome. The eastward enlargement of the European Union in May 2004 aims to ensure peace, security, prosperity and stability on the continent. Moreover, in many Central and Eastern European countries, a new and modern knowledge-based society is replacing the traditional economic model based on industrial production. This opens up new socio-economic perspectives and ensures that these countries will be potent players and competitors in the European economy.

At the same time European integration poses a serious challenge. In Central and Eastern Europe, the transformation from totalitarian states and centrally-planned economies to western democracies and market economies has caused dramatic shifts in economic, social and ecological development. It became clear at an early stage that this transformation has a spatial dimension. Thus, while some regions have been able to take advantage of the new situation, others have been seriously weakened by structural change. Many cities and regions which prospered during an era when political priorities were focused on industrial development have suffered rapid economic decline. Unemployment, social problems and environmental damage were some of the consequences which now place a heavy burden on the development of these areas.

In the late nineties the Dresden-based Leibniz Institute of Ecological and Regional Development (IOER) initiated a research project on the future of industrialised cities and regions undergoing structural change (FOCUS). It was conducted in the framework of the Network of Spatial Research Institutions in Central and Eastern Europe, coordinated by the IOER. The objectives of the project were to analyse the problems of transformation in affected regions, to compare the strategies which have been or are currently applied in order to deal with the problems, and to initiate a wider discussion on the challenges related to the topic.

With the support of the EU Community Initiative INTERREG II C, designed to encourage interregional cooperation and the exchange of experience, research teams from eleven countries were involved in the project: Austria, Bulgaria, the Czech Republic, Germany, Hungary, Latvia, Poland, Romania, Slovakia, Slovenia and the Ukraine. In 2000, the Leipzig Declaration of the Network of Spatial Research Institutions in Central and Eastern Europe became an important milestone in international discussion about the future of industrialised cities and regions undergoing structural change.

This book highlights the major results of the FOCUS project which from 1999 ran over several years and in the mean time has resulted in concrete pilot projects in a number of Central and Eastern European countries. Eleven country reports are presented. They largely refer to the situation in the late nineties and the beginning of the new century, and are placed into the framework of the current state of discussion.

Apart from thanking the EU INTERREG II C programme, we would also like to express our gratitude to the German Federal Ministry of Transport, Building and Housing (BMVBW) and the Austrian Federal Ministry of Education, Science and Culture for co-financing the project. Our special thanks go to André Müller and Jens Kurnol of the Federal Office for Building and Regional Planning and Welf Selke of the BMVBW who were committed in support of the FOCUS project.

We also wish to thank all the colleagues from the other research institutions with whom we shared an enriching work experience, a challenging exchange of ideas and the possibility to learn from one another. We are grateful to Tassilo Herrschel of the University of Westminster, London, for his constructive comments on the current state of the international discussion. Last but not least we would like to thank Karina Pallagst and Karl Schmude for their help in managing the complex international and interdisciplinary project, Sabine Witschas and Katrin Kettner for designing the maps and creating the FOCUS internet homepage, Margitta Wahl for helping to prepare the manuscript, Lars Liebe, Kerstin Röhner and Katharina Mörl for their support of the organisational and scientific work, as well as Derek Henderson for his English language editing.

Dresden and Bratislava, May 2004

Bernhard Müller, Maroš Finka and Gerd Lintz

Contents

Introduction: The Challenge of Structural Change for Industrial Cities and Regions in the CEE Countries

Gerd Lintz[1], Bernhard Müller[2] and Maroš Finka[3]

[1] Leibniz Institute of Ecological and Regional Development (IOER), Dresden
[2] Leibniz Institute of Ecological and Regional Development (IOER), Dresden, and Dresden University of Technology (TU Dresden)
[3] Slovak University of Technology, Faculty of Architecture, Bratislava

1 What About the Losers? Background and Aim of the Book

1.1 Origin, Aim and Structure of the Book

With the fall of the iron curtain, the unprecedented transformation from socialism to western democracy and market economy in the post-socialist Central and Eastern European (CEE) countries at the beginning of the 1990s entailed fundamental economic, social and environmental changes in the European transition countries, and catalysed an acceleration in the processes of post-industrial transformation all over Europe. The spatial impact of the post-socialist transformation process in general has been addressed in a plethora of publications (e.g. Hamilton 1995, Smith 1996, Keune 1998, Pütz 1999, Bachtler, Downes and Gorzelak 2000, Gorzelak et al. 2001, Mayr, Grimm and Buchholz 2001). Generally, the authors state that industrial cities and regions in particular are detrimentally affected by the changes; they are even termed "the losers" of transition (Gorzelak 1998, see also Förster 1999 and 2000, Kunze 1997).

At the same time there seems to be a rather limited awareness of the situation in old industrial cities and regions in the CEE countries. This seems somewhat surprising given the high profile such problems were accorded by policymakers some 10-15 years earlier; it seems that less attention is currently being paid than when Western Europe itself was struck by strongly regionalised structural change (see e.g. Hamm and Wienert 1990, Häußermann 1992). There exist very few studies investigating the situation of these cities and regions across the level of the CEE countries (e.g. rudimentary analyses by Kwiatkowski, Janusz and Steiner 1995, Fazekas and Gorzelak 1995, Kunze 1997).

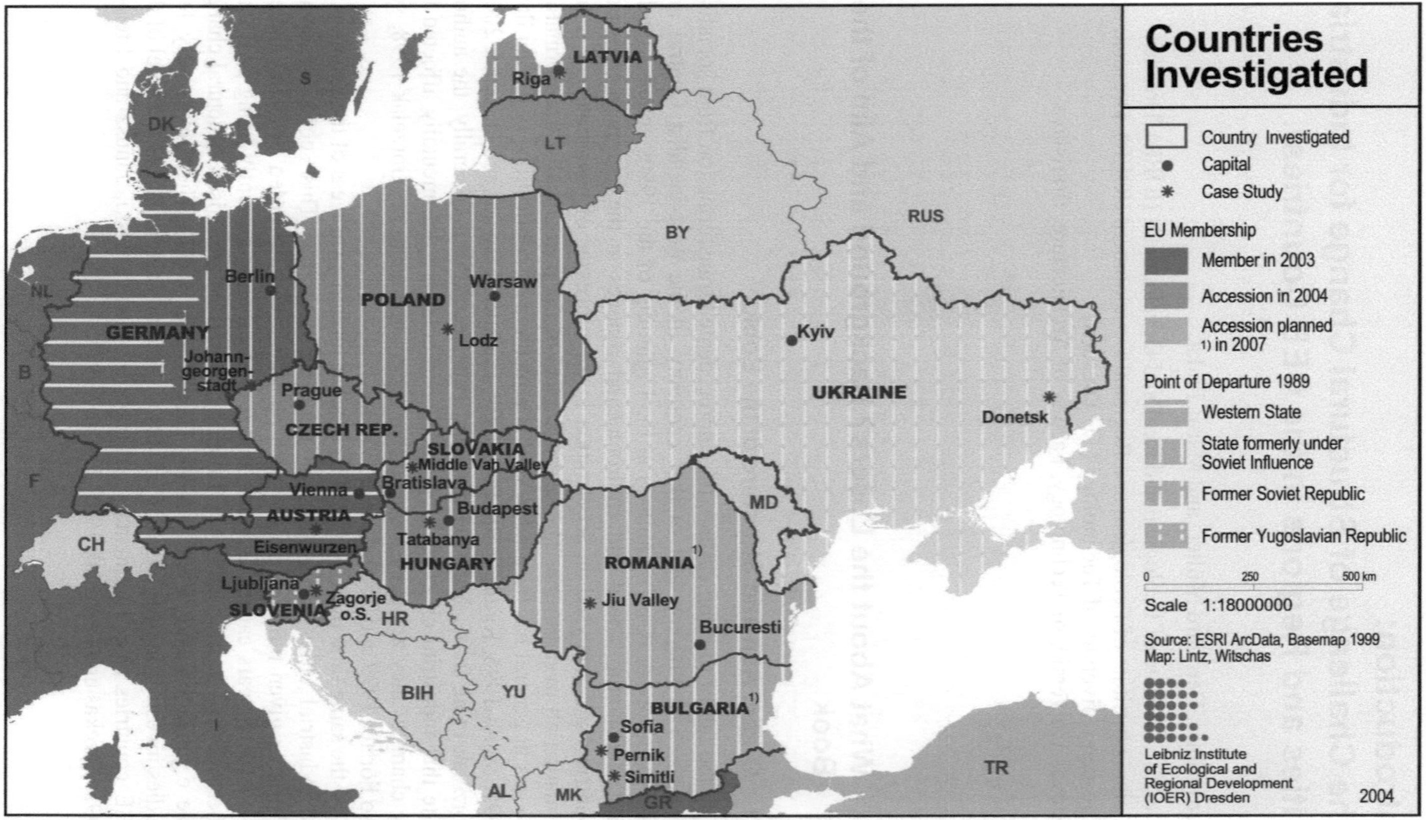

Map: The investigated countries

Numerous cities and regions which for decades had been the motors of national economic development owing to their strong industrial base, are now confronted by considerable socio-economic and environmental problems in the wake of the sudden collapse.

The adjustment processes in the former economic heartlands of the socialist states have had a severe impact, not only economically and socially, but also environmentally. Under socialism, economic production targets were paramount, while environmental costs were seen as a necessary price to be paid. The state stifled any criticism to avoid public debate of the issues (see e.g. Herrschel and Forsyth 2001).

The problems faced by these old industrial cities and regions in the CEE countries, particularly in transition CEE countries, were at the centre of a pan-European research project entitled "The Future of Industrialised Cities and Regions Undergoing Structural Change (FOCUS)", partly funded by the EU INTERREG IIC programme and which ran over two years from 1999 to 2001 (see www.ioer.de/FOCUS/home page.html).[1]

The aims of this project were to investigate and compare the specific problems of cities and regions dominated by old industries, to explore the strategies that have been adopted to cope with the structural changes triggered by the end of the socialist systems, and to initiate a Europe-wide discussion about this topic. These comparisons, it was envisaged, would allow clear identification of the different circumstances, and the likely effects of local and regional solutions to the common problem of industrial collapse in the face of new global competition. Furthermore, it was hoped that evidence of possible lessons to be learned about "good practice" in designing effective policies could be garnered, especially in view of EU expansion. According to the funding stipulations of the INTERREG IIC programme, a direct exchange of knowledge and experience between the project partners and other interested parties/actors should take place.

The project covers nine CEE countries, plus the "old" EU member states Germany and Austria (see Map). The CEE countries include the former satellite states of the Soviet Union – Bulgaria, the Czech Republic, Hungary, Poland, Romania and Slovakia, the former Yugoslavian Republic Slovenia and the former Soviet Republics of Latvia and Ukraine. Of these case studies, almost all CEE countries joined the European Union on May 2004; the exceptions were Bulgaria and Romania, which are expected to join in 2007, and the Ukraine, which is currently not a candidate for accession.

Thus, countries with very different political and economic legacies are involved, and there is a distinct east-west gradient in terms of proximity to the EU, both politically and economically. This may correspond with pressure to adopt EU principles of governance. The result has been a relatively detailed picture of adjustment processes and underlying problems, highlighting the diverse range of policies necessary to effectively address post-socialist economic development in

[1] The project was funded by the EU community initiative INTERREG II C, the German Federal Ministry for Transportation, Building and Housing (BMVBW), the Austrian Federal Chancellery and the participating institutions.

old industrial regions in the CEE countries. The emerging picture of the relevant coping strategies, by contrast, is still somewhat blurred. A general trend seems to be the use of a wide variety of strategies, influenced by the general approach to transition and the EU accession process.

The project included regular discussions of the eleven project partners, these agreeing on a basic common structure to the countries' reports on the problems and coping strategies of the industrial cities and regions. The emerging results were broadly debated at two conferences organised by the Network of Spatial Research Institutions in Central and Eastern Europe (see www.ioer.de/cee-net/homepage.htm) in Krakow and Leipzig respectively, also involving young researchers who contributed urban and regional case studies.

This book seeks to document the results of the project by presenting the eleven country reports (see also Lintz, Müller and Schmude 2004). Case studies about single cities and regions are included in these reports as boxed inserts. The introduction of the book outlines a conceptual basis and points out some important framework conditions under which the industrial cities and regions have developed. Firstly, the transition process and its impact on the industrial cities and regions must be analysed. Since European integration is an important goal of the CEE countries and an important factor of influence on strategies implemented, the introduction deals, secondly, with EU enlargement. In the last section the countries are compared more broadly using socio-economic indicators. The book concludes with a summary of the results and with recommendations for future policymaking.

1.2 Industrial Regions as Problem Regions

Analysing spatial development in the wake of transformation, Gorzelak (1998 and 2002) identified four different types of regions by using two criteria: the position in the socialist economy and the reaction to transformation (see Table 1). The "leaders" and the "losers" shall be discussed here.

Metropolises and Capitals are termed leaders in view of their diversified economy, skilled labour-force, good infrastructure and rich institutions. They are characterised by a previously good position in the socialist economy and a positive reaction to transformation. One can easily argue that, given the general emphasis on industry and heavy industry in the socialist economy, many of them could also have been termed industrial cities or regions. However favourable conditions have allowed them to make great progress in modernising and restructuring their industries and societies. They have managed to counter the decline of affected industries by creating new competitive jobs in industry or in the tertiary sector.

According to Gorzelak, the losers in spatial development are the industrial cities and regions. In contrast to the leaders, they reacted negatively to transformation and are characterised by a specialised industry, derelict land and biased qualification. So, taking up the idea that industrial cities and regions could cope differently with the challenges of transformation, the losers are those with unfavourable conditions for development. They certainly form the majority of those who in socialist time could be called industrialised.

Table 1. Regional patterns of transformation

		Reaction to transformation	
		Positive	Negative
Position in the socialist economy	Good	LEADERS Positive continuity Metropolises and capitals Diversified economy, skilled labour, good infrastructure and rich institutions	LOSERS Negative discontinuity Industrial regions Specialised industry, derelict land, biased qualifications
	Bad	WINNERS Positive discontinuity Tourist and border regions Diversified economy, external demand	LAGGARDS Negative continuity Rural, peripheral Nonaccessible, obsolete structure, low qualifications

Source: Gorzelak 1998 and 2002, slightly modified

This book focuses on the losers, i.e. those industrial cities and regions which show severe problems arising from structural change, and follows roughly the definition used by the European Union's Structural Funds (objective 2 regions 1989-1999) to characterise "regions affected by industrial decline" (cf. Vanhove 1999, 491f.):

- the percentage of jobs in the industrial sector is higher than average, characterising the region as industrial;
- the unemployment rate is above average, marking the region as a problem region;
- there is a decline in industrial employment, indicating industrial decline as the main cause for high unemployment.

Sometimes the cities and regions under investigation are also termed old industrial/industrialised areas or industrial problem areas. Often the decline of industry cities and regions is so severe that they can in fact no longer be termed industrial. Therefore our use of the term "industrial" refers chiefly to the state which prevailed before structural change.

The fact that regions of industrial decline exist presents a challenge for the ongoing process of European integration, in view of the fundamental European aim of social and economic cohesion. Consequently, the European Spatial Development Perspective (ESDP; Informal meeting of the ministers for spatial development 1998) sees the restructuring process of industrialised cities and regions as a problem of European dimension. On the other hand, the same document highlights the possibilities of economic and urban renewal.

2 Transition: the Challenge for Industrial Cities and Regions

Western European market-oriented countries have seen a decline in their industrial cities and regions since at least the sixties of the twentieth century. In Western Germany, for instance, the coal mining sector in the Ruhr area began losing importance at that time. The problems in such regions have been the subject of public debate and have been examined in a number of publications (e.g. Hesse 1988, Hamm and Wienert 1990, Häußermann 1992, Schrader 1993, Wood 1994, Cooke 1995, Blotevogel 1997).

In socialist countries the phenomenon of declining industrial cities and regions hardly existed before 1990 (with some exceptions, e.g. uranium mining).Sectoral production and the spatial structure of the economy were centrally planned by national governments and unemployment was artificially avoided. The radical change in the CEE countries which started in 1989/1990 led to an unprecedented transition in political, social and economic terms. This transition and, in particular, the consequences for industrial cities and regions have not been thoroughly analysed.

In the first section, the political, social and economic dimensions of transition are outlined. The second section deals with the structural economic changes which are triggered in the process of transition. The challenge for industrialised cities and regions of successfully handling structural change is examined in the last section.

2.1 The Political, Social and Economic Dimension of Transition

Since 1989 Europe has experienced a unique peaceful transformation or transition (e.g. Gros and Steinherr 1996, Pickles and Smith 1998, Roland 2001). Crampton describes the dramatic early years as follows: "In the late summer and early autumn of 1989 waves of reform and revolution crashed upon eastern Europe's communist régimes with such a force that within two years the entire system had been swept away" (2000, 391). The transformation has taken and still takes place in the political, economic and social dimension (Stadelbauer 2000). Politically, the rule of law has been clarified and a broader participation of the population introduced, leading to significant changes in the structure of the political-administrative system and a strengthening of democracy.

The new political freedom has been accompanied by economic freedom. As the enthusiastic words of Václav Havel from 1993 show, critics of the former socialist countries were convinced that there had to be a fundamental change in the economic system: "A market economy ... is the only natural economy, the only kind that makes sense, the only one that can lead to prosperity, because it is the only one that reflects the nature of life itself. The essence of life is infinitely and mysteriously multiform, and therefore it cannot be contained or planned for, in its fullness and variability, by any central intelligence" (after Samuelson and Nordhaus 1998, 265). According to Kloten (1993, 124f.) the policy of economic transformation ideally involves four stages:

- creation of a new legislative and institutional framework, e.g. code of private law, privatisation, banking system, public budget and tax system, organisation of the labour market, unemployment, old age and health insurance;
- stabilisation of the overall development of the economy, budgetary consolidation, development of money and capital markets, freeing prices in the goods and factor markets, opening-up markets to the world economy, social policy to cushion the hardships of transition;
- economic policy for adjustment of structures in the real sphere of sales, of trade and payment flows, of relative prices, employment etc.; privatisation on a broad front, creation of conditions for the creation of innovative processes;
- systematic development of a free market system that is both efficient and equitable.

The biggest difficulties probably arise in connection with the social dimension of transformation. And indeed, in contrast to the ideals of socialism, the new societies evolving in the CEE countries are highly stratified: there is a growing gap between rich and poor, between employed and unemployed. Emerging pluralism has sometimes led to a state of uncertainty. Broadly one can say that a civil society is arising.

It is important to note that according to their historical legacies and general basic conditions the CEE countries have adopted various speeds of transition and focused on different aims (cf. e.g. Dostál 1998). Countries such as the Czech Republic, Hungary, Latvia, Poland and Slovakia decided to endure "shock therapy", while Romania and Ukraine preferred a rather gradual transition. The German Democratic Republic speedily joined the Federal Republic of Germany and immediately adopted its "western" laws and regulations.

2.2 Transition and Economic Structural Change

Industrial structure in market economies is in principle determined by factors on the demand (consumers) and supply sides (producers) (cf. Pindyck and Rubinfeld 1998). Continual shifts in consumer demand are closely linked to developments in income. On the supply-side, the availability of production factors and the degree of innovation in the field of products and production processes are the most important factors. Innovations in industrial production processes, for example, lead to jobs losses in that sector if the increase of productivity can not be compensated by growing sales. If additionally the demand for services rises with increasing incomes while the productivity of the service sector increases only slowly, then the share of services in GDP and employment rises (three sector hypothesis).

Globalisation has an even bigger impact than changes in domestic factors, and can quickly affect the national and regional conditions of supply and demand. The steady reduction of barriers to trade and capital flows (decrease of tariffs, transportation and communication costs), as for example initiated by the World Trade

Organisation or the European Union, leads to a new international division of labour which puts a tremendous pressure on national economic structures. Firms respond to this accelerating pressure by changing their way of organising production. The fordistic pattern of production, with large plants and vast unit production becomes less attractive, and instead the flexible production of smaller numbers of produced units becomes more important. Co-operation between firms increases and the intensity of co-operation grows (Murray 1992, Butzin 1993, Hudson 1997).

In contrast to the western market economies, the economic structures in the former socialist countries, e.g. the contribution of industries to the GDP or the organisation of production, were controlled by socialist central government planning and integrated within the Council for Mutual Economic Assistance (COMECON). This led to a completely different set of aims. Subsequently economic structural development deviated strongly from the structures which would have emerged on an open market economy, and the adaptations necessary to function on the globalised world economy have been immense. Not only were the economic structures different, but also a lack of innovation and investment resulted in poor overall productivity and an inferior quality of goods (Werner 1995, Gros and Steinherr 1996). The deficient innovation and investment also caused serious environmental pollution of air, water and soil (Tickle and Welsh 1998, Carter and Turnock 2001).

The socialist economy placed emphasis on mass production and heavy industries, organised within big state-owned conglomerates with a relatively narrow range of production. With the opening of their economies, the transition countries were able to buy investment and consumer goods on the world market and attract foreign direct investment. However their industries also faced fierce competition from foreign competitors. In particular the break-up of COMECON forced companies to reorient their purchasing and selling practices.

Thus the former socialist countries have had to bridge the gap between their old structures and the prevailing structures in the world economy as well as adapting to the new pace of structural change as determined by globalisation processes. This adaptation was accelerated in comparison to Western countries, and the transition much more difficult than expected. Economic activity fell in all transition countries, at least for some years, with reduced GDP (cf. chapter 1.4). The contraction of industry was particularly striking.

2.3 The Challenge for Industrial Cities and Regions

The core question concerning the challenge of transition for industrial cities and regions is: What does the process of transition and the ensuing pressure for economic structural change mean to the industrial cities and regions? As already mentioned there are numerous publications about industrial problem regions in general, mainly examining the situation in western countries. Different theories have been posited to try to explain the reasons for the decline of industrial cities and regions in market economies and give a basis for the development of useful strategies (Steiner 1985, Butzin 1987, Tichy 1987, Wößmann 2001). There is no attempt to present the different theories in this book. Instead a rather pragmatic approach has been adopted, integrating the most important aspects.

A starting point is to isolate the three main determinants of development in industrial cities and regions: (1) the vulnerability of the local or regional economy to changes in framework conditions, such as growing competition or shrinking markets in certain branches of industry, (2) the ability of the local and regional economy to react positively to changes and (3) the ability of the local and regional political-administrative system to support the economy and the adjustment process.

(1) *Vulnerability towards changes in the framework conditions.* Often the decline of industrial cities and regions is attributed to particular industries that are negatively affected by changes of the framework conditions, such as the emergence of new competitors in coal mining or steel industry. Certainly, the bigger the regional share of those industry sectors under pressure the bigger the pressure placed on the whole regional economy to adapt, and the worse the prospects of achieving sustainable economic development. Therefore, affected single-industry regions face the most difficult situation.

When, in western countries, selected industries lost their international competitiveness, cities and regions with a particular concentration of firms or jobs in, say, coal mining, the steel industry, ship building and textile industry faced structural change and possible decline. Other cities and regions, often abroad, could take advantage of the change in pattern. However in the former socialist countries the situation is rather different: on the one hand certain industries are under a similar pressure to that faced in western countries. But on the other hand, almost all industries have had to adapt to the new circumstances to a greater or lesser degree. Which cities and regions suffer decline and which cities and regions can prosper becomes only apparent over time.

Studies show that often affected industrial sectors in old industrial regions perform worse than those affected industrial sectors in other, more successful regions (Steiner 1985, Häußermann 1992). This means that there must be additional local or regional reasons for the decline of these sectors and the poor situation of the local and regional economy, beyond the simple share-of-industries approach.

(2) *Ability of economy to adapt positively.* The second element, the ability of the economy to react positively to changes, touches on two questions: How does the affected industry deal with its problems, and, what can the other sectors contribute to alleviating job losses? In general, unemployment in cities and regions facing adaption pressures can be avoided in two ways (cf. Cooke 1995). Either the affected companies can maintain employment by diversifying the product range or finding new market niches (internal structural change), or other companies can create new sustainable jobs to offset job losses in the "old" industries (external structural change).

Before the local and regional factors which support the establishment and growth of companies is discussed, the context of transition and the basic national conditions for a growing market economy should be mentioned again. The state of transition and the success in coping with the problems of transition generally determine the likelihood of coping with structural change. For instance, insufficient privatisation, a high inflation rate, legal uncertainty, criminality and corruption can undermine the efforts of transition over the whole country.

As generally the market economy is highly dependent on the quality of management and entrepreneurs, positive internal and external economic structural change relies very much on the capability of local and regional managers and entrepreneurs. They set up businesses, run small and medium-sized enterprises or develop businesses into large firms. They are responsible for estimating demand, for developing new products and the reduction of costs. Thus they are the key persons to ensure income growth and rising employment. Cities and regions must have good managers and entrepreneurs, as they are the endogenous motors of economic development. Entrepreneurial thinking and knowledge, and a willingness to innovate and take risk is heavily influenced by regional history, culture and tradition. In this respect the former socialist countries are generally at a serious disadvantage. The domination of large plants or combines (Grabher 1997) often leads to a particular local or regional lack of entrepreneurship (Hamm 1991, Wood 1994).

The success of managers and entrepreneurs in developing their companies greatly depends on the local and regional framework conditions for economic development. These location factors are also important for foreign companies considering investing in the industrial problem cities and regions (Blakely and Bradshaw 2002, Molle and Verkennis 1995, Hodgkinson, Nyland and Pomfret 2001). In the following some location factors relevant for industrial cities and regions are outlined (cf. Butzin 1987, Hamm 1991, Cook 1995, Förster 1996, Wood 1997):

- The supply of labour is important for locational decisions. As a rule industrial cities have a pool of skilled and unskilled workers specialized in the previously prevailing industries. Wages are often relatively high. New industries and services normally require quite different, often higher qualifications.
- The problem with available infrastructure is similar to the problem of labour supply. Infrastructures were designed to support the old industries. The decline in importance of railways and harbours, for example, has corresponded with increasing reliance on motorways, airports and, especially, communication networks. New firms and industries need appropriate buildings and plants.
- The redevelopment of old commercial and industrial zones, often helpful in improving settlement structures, is often too expensive in view of such negative factors such as contamination. Cities and regions need additional space not always available in agglomerations, and this lack of space can hinder the economic development.
- More and more importance is being placed on the attractiveness of the urban and regional environment (e.g. quality of air, water and countryside) (Grabow, Henckel and Hollbach-Grömig 1995). Well qualified workers and managers, in particular, value an attractive environment. Industrial cities and regions often lack such "soft" locational factors, particularly in the CEE countries (cf. Pavlínek and Pickles 2000). This affects both cultural attractiveness and other fac-

tors governing quality of life. Often smaller mining towns are situated in an attractive natural environment, offering some potential for tourism.

- Institutions such as universities and research institutions are seen as increasingly important. They provide a highly qualified labour-force and potential entrepreneurs while also enhancing the regional innovation capacity. Often in industrial cities and regions there is a dearth of general higher education facilities and general research institutions.
- In this context the importance of co-operation between universities, research institutions and businesses should also be highlighted. Such regional innovation networks can only emerge when there is a culture of co-operation, an "innovative milieu". These "ultra soft" locational factors require – like some other locational factors – a degree of agglomeration and often need a certain structure of industries in place upon which to build (new) industrial clusters (Cooke 1995, Sadler 2004). Some industrial cities and regions, e.g. capitals, already have the requisite basis to exploit this factor to ensure sustainable jobs and high incomes. In (old) industrial cities and regions structural change is often hampered rather than helped by existing networks (Grabher 1993).
- Lastly, it should be pointed out that the spatial location of industry in the CEE countries was largely unconstrained by the logic of the market economy. Socialist spatial development policy aimed at industrialising even rural and peripheral regions, placing a heavy burden on industrial development in these areas.

(3) *Ability of the political-administrative system to support structural adaptation.* Many of the location factors are determined or supported by public policy, both at a local and regional level. Even in market economies governments are supposed to encourage economic development, or at least not to hamper it without good reason. So the ability of the local or regional economy to adapt depends very much on the ability of the political-administrative system to support the economy and the adjustment process, even becoming part of the innovation network.

The quality of governments (politicians, personnel and organisational structure of administration, readiness for change, experience) and the co-ordination and co-operation between the government, businesses and, not least, citizens can thus be viewed as a separate location factor. Policy-makers need knowledge, experience and openness to innovation in much the same way as entrepreneurs. They have to improve their location potential in the most efficient way. Since a city or region cannot hope to attract all kinds of industries, the industry/sector (new industries, services, mixture) with the most favourable long-term prospects must be assessed according to the regional characteristics and factor endowment. Often in the transition countries, and especially in the old industrial cities and regions, policy-makers still have insufficient knowledge, experience and openness to innovation.

Shortcomings in one of the determinants of development or location factors can establish or worsen weaknesses in other factors. As already implied, the problems of industrial cities and regions are often interwoven and interdependent, similar to the cumulative causation model for underdeveloped regions by Myrdal (1957). There is a vicious circle of decline in industrial output and loss of image and self-esteem among the population, which in turn leads to an exodus of urgently needed highly-qualified and youthful workers. These are the very people needed for the development of new economic structures, particularly in those areas unattractive to external corporations. As a result, local and regional public bodies are financially overwhelmed by the maintenance costs of massive infrastructural legacies, including derelict factories and the associated clean-up costs of contaminated sites. These problems, together with the growing indirect costs of social disintegration and poor economic perspectives, all combine to form a complex set of problems shaped by particular local and regional conditions.

Conversely, one positive determinant or location factor can compensate other negative factors, and improvements in one field can kick-start additional improvements in another field (Blakely and Bradshaw 2002).

3 Accession to the European Union

EU accession of the CEE countries is of special importance for the transformation process of old industrialised cities and regions. While structural and political change began abruptly and furiously in most CEE countries in 1989/1990, the idea of joining the European Union matured only slowly. But once wholeheartedly accepted, the influence of the accession process on the shaping of the transformation accelerated and took a dominant position (cf. Berglöf and Roland 1997). Only fifteen years after the fall of the iron curtain, the old divisions of Europe were overcome on the 1st of May 2004 when ten countries from Central and Eastern Europe and the Mediterranean joined the EU. Included were six countries investigated in this book: the Czech Republic, Hungary, Latvia, Poland, Slovakia and Slovenia. Already in 1990 the German Democratic Republic had joined the EU by merging with the Federal Republic of Germany, while Bulgaria and Romania plan to join the EU in 2007. Currently the Ukraine is not seen as a candidate state.

EU enlargement has been regarded as a historic opportunity by both old and new members to bring freedom, political stability, peace and enhanced economic prosperity to Europe. From the perspective of the accession states, EU membership also poses a historical challenge: to quickly absorb the full range of EU rules, practices and presumptions, usually referred to as the "Aquis Communautaire", which has developed over 50 years with no influence from the new members. The challenge facing the enlarged EU is to develop workable political structures to accommodate twenty-five members, while supporting the accession states in their efforts to adapt to the EU model and to catch up in terms of economic, social and environmental standards.

Before a brief overview of the Aquis Communautaire is given, a short "history" of EU enlargement will give some background to the most recent events. Finally, the general framework of EU assistance is discussed, with particular regard to the accession candidates.

3.1 Short "History" of Eastward Enlargement

After the "silent" enlargement of the EU through German reunification in 1990, three years went by before the European Council in Copenhagen officially stated the possibility of the CEE countries joining the EU (cf. European Commission 2001, EU website for enlargement[2] and www.euractiv.com). The following preconditions, the so-called "Copenhagen criteria", were formulated: "stability of institutions guaranteeing democracy, the rule of law, regard for human rights and respect for and protection of minorities; the existence of a functioning market economy as well as the capacity to cope with competitive pressure and market forces within the Union; the ability to meet the obligations of membership, including adherence to the aims of political, economic and monetary union" (European Commission 2001).

Poland and Hungary submitted their application for membership in 1994, with most of the other CEE countries applying in 1995 and the Czech Republic and Slovenia following one year later. In 1995 the Madrid European Council highlighted the necessity of candidate countries transforming their administrative and judicial structures so that EU law could be effectively implemented at national level.

In 1998 accession negotiations were initiated with the Czech Republic, Hungary, Poland and Slovenia. Negotiations began one year later with Latvia, Bulgaria, Romania and the Slovak Republic. The Helsinki European Council in 1999 declared that the accession countries "must share the values and objectives of the European Union as set out in the treaties"; it also stated that the candidates should resolve outstanding border disputes, and stressed the importance of high standards of nuclear safety.

In the 2000 Nice European Council the candidate countries were urged to "continue and speed up the necessary reforms to prepare themselves for accession" (European Commission 2001). At the second Copenhagen Summit in 2002 accession negotiations were concluded with those countries joining the EU in May 2004; the Treaty of Accession was signed at the European Council in Athens in 2003.

3.2 The Challenge: Adoption of the Aquis Communautaire

Accession negotiations deal with the terms under which the candidate states adopt, implement and enforce the Aquis Communautaire (EU website for enlargement). The negotiations are structured over thirty-one chapters, relating to matters of community competence. The following chapters have most relevance to the problems of industrial cities and regions and strategies for their recovery.

Chapter 1 to 4 refer to the removal of barriers which hamper the free movement of goods, persons, capital and the freedom to provide services within the EU. This

[2] EU website for enlargement: http://europa.eu.int/comm/enlargement/index_en.html, May 2004.

leads to stronger sales competition between companies and stronger competition between cities and regions for qualified workers and capital. Competition policy is dealt with in Chapter 6. One feature of EU competition policy is to reduce the practice of using subsidies to support individual companies in designated areas. Chapter 15 concerns EU guidelines for industrial policy. Candidate countries are requested to present their industry policy and restructuring strategies in order to assess whether these policies conform to the EU principles.

In Chapter 21 the preparations necessary to adopt EU regional policy are set out. By the time of accession the following provisions for the implementation of EU regional policy must be completed: the appropriate legal framework, a territorial organisation according to the NUTS classification (Nomenclature of Territorial Units for Statistics), appropriate programming and administrative capacity, and sound financial and budgetary management. The new members can only profit from the EU structural funds when the appropriate system is in place.

Chapter 22 treats EU environmental legislation. This covers "environmental quality protection, polluting and other activities, production processes, procedures and procedural rights as well as products. Apart from horizontal issues (environmental impact assessments, access to information on environment, combating climate change), quality standards are set for Air, Waste management, Water, Nature protection, Industrial pollution control, Chemicals and genetically modified organisms, Noise and Nuclear Safety and Radiation Protection" (EU website for enlargement).

3.3 EU Assistance: PHARE, ISPA, SAPARD and TACIS

The background: EU assistance for the (old) EU member states

According to the Treaty of Amsterdam (1997), which amended the Treaty on the European Union (1992), the Union aims at an overall harmonious development, and in particular looks to easing developmental disparities between the various regions (Article 158) (Vanhove 1999, 419). The primary means for achieving this aim in the framework of EU regional policy are the EU Structural Funds, to which about one third of the EU budget is allocated. The European Regional Development Fund (ERDF) is the most important of these Structural Funds.

In the programming periods 1989-1993 and 1994-1999 the following regions were eligible for support (Vanhove 1999):

- regions whose development is lagging behind (Objective 1),
- regions seriously affected by industrial decline (Objective 2),
- rural areas (Objective 5b) and
- regions with an extremely low population density (Objective 6, since1994).

In the third programming period 2000-2006 the number of objectives has been reduced. Most regions previously eligible as objective 6 regions are now categorised under the new objective 1. The new objective 2 regions – areas undergoing economic and social conversion – encompass the former Objective 2 and Objective 5b regions.

Apart from these regional objectives, the Structural Funds also finance *Community Initiatives*. In the programming period 1994-1999 about 9 percent of expenditure from the Structural Funds was allocated to 13 initiatives, some of which dealt explicitly with the diversification of activity in those industrial areas heavily dependent on failing sectors. These Community Initiatives were "Resider" for steel regions, "Renaval" for shipbuilding regions, "Rechar" for coal-mining regions and "Retex" for textile regions. In the programming period 2000-2006 these initiatives were abolished, but the aim of diversification can be realised – on a much smaller scale – within the four community initiatives still running (INTERREG, URBAN, LEADER and EQUAL) (EU website for regional policy[3]).

Finally, two of the main principles of EU Regional Policy should be mentioned. According to the first principle the EU finances multi-project *programmes* running over several years. They are supposedly based on a sound regional analysis and include a clear strategy to tackle regional problems (Armstrong and Taylor 2000, 330). The second principle is that of *partnership and subsidiarity*, meaning the aim of ensuring a close collaboration between the Commission and all relevant authorities at national, regional and local level at all programme stages.

Separate from the Structural Funds, a Cohesion Fund was established in 1993. This Fund is not regionally oriented and finances projects designed to improve the environment and develop transport infrastructure in member states whose per capita GNP is below 90 % of the Community average (i.e. Greece, Ireland, Portugal and Spain).

Regional policy will be revised at the end of the programming period 2000-2006, taking closer account of eastward enlargement and the new economic disparities within the EU (cf. European Commission 2004).

EU assistance for the accession states

Already in 1990 the EU started the PHARE programme to support the former socialist CEE countries in their efforts to cope with the difficulties of transformation. This support evolved into pre-accession aid (cf. EU websites for enlargement and regional policy, Sajdik 2004, Vanhove 1999). EU assistance for the accession countries could in part be used to support the rehabilitation process in old industrial cities and regions. The assistance includes three main instruments: PHARE, ISPA and SAPARD.

The *PHARE* programme (Poland and Hungary Action for the Reconstruction of the Economy) applies to the acceding and candidate countries from Central and Eastern Europe, and principally involves measures to build institutions (with accompanying investment), as well as measures designed to promote economic and social cohesion. In the last few years the PHARE programme has concentrated on reinforcing the administrative and judicial capacity of applicant states and easing adoption of the Aquis Communautaire. In line with the programme's

---

[3] EU website for regional policy: http://europa.eu.int/comm/regional_policy/index_en.htm, May 2004.

various aims, a number of sub-programmes have been introduced. These include STRUDER, which supported integrated development programmes in regions with restructuring problems, and COOPME, which assisted small and medium-sized enterprises.

The *ISPA* programme (Instrument for Structural Policies for Pre-Accession) is similar to the Cohesion Fund. It is targeted at environmental measures and improved transport infrastructure (connections between CEECs and the trans-European networks). The *SAPARD* programme (Special Accession Programme for Agriculture and Rural Development) focuses on agriculture and rural development. ISPA and SAPARD were both established in 2000.

With their accession to the EU in May 2004 the ten new members became eligible to receive aid from the Structural Funds and the Cohesion Fund. For the period 2004 to 2006 special arrangements of financing and assistance have been set up; as of 2007 the new member states will be fully integrated within a revised EU regional policy. Bulgaria and Romania will continue to receive pre-accession aid till their accession to the EU, planned for 2007.

Apart from pre-accession aid, the EU also supports countries in Eastern Europe and Central Asia, including the Ukraine, with the TACIS programme (Technical Assistance to the Commonwealth of Independent States). It aims at enhancing the transition process in these countries (EU website for External Relations[4]).

4 The Investigated Countries: Basic Information

The map given at the beginning shows the diverse nature of the countries investigated in this book. They vary considerably in size and topography, and historically they had different political points of departure in 1989 withdiverse attitudes and relations to EU membership.

Concerning the *size* of the countries, Table 2 gives further information. The Ukraine is by far the biggest country at around 600,000 square kilometres, followed by Germany (357,000 km^2) and Poland (312,000 km^2). The smallest country is Slovenia with around 20,000 square kilometres, the second and third smallest countries are Slovakia (49,000 km^2) and Latvia (64,000 km^2). The number of *inhabitants* in not exactly proportional to the size of the countries. Germany shows the biggest population with 82 million inhabitants, while the Ukraine has 49 million and Poland has 39 million inhabitants. Slovenia shows the smallest population with 2 million inhabitants, with Latvia 2.4 million and Slovakia 5.4 million.

A comparison of the *gross domestic product (GDP) per capita* highlights the former division of Europe into two blocs with contrasting economic systems and different levels of prosperity. Austria and Germany have a long tradition as free market economies and show GDPs per capita of around 25,000 and 24,000 US Dollars (USD) respectively. GDP per capita in the CEE countries varies signifi-

---

[4] EU website for external relations: http://europa.eu.int/comm/external_relations/ceeca/index.htm.

cantly and seems to reflect the timetable of accession to the EU. Slovenia is the economically strongest country with GDP per capita of 11,000 USD. The Czech Republic and Hungary both have a GDP per capita of around 7,000 USD.

Table 2. Basic data of investigated countries

	Surface (th km²)[1]	Total population (million) 2001[2]	GDP per capita, 2002, current prices (USD)[4]	Unemployment rate, 2001[2]
Austria	83.9	8.1	25,131	3.6
Bulgaria	110.9	8.0	1,932	19.2
Czech Rep.	78.9	10.3	6,784	8.0
Germany	357.1	82.3	24,197	7.7
Hungary	93.1	10.0	6,646	5.6
Latvia	64.6	2.4	3,581	12.8
Poland	312.7	38.7	4,884	18.5
Romania	237.5	22.4	2,046	6.6
Slovak Rep.	49.1	5.4	4,389	19.4
Slovenia	20.3	2.0	11,031	5.8
Ukraine	603.7	49.3	856	(3)11.1

1 Source: European Commission, Reports of candidate countries 2003
2 Source: United Nations Development Program, HDR 2001
3 Source: United Nations Economic Division for Europe, Statistics 2003
4 Source: IMF 2003

A second group comprising Poland, Slovakia and Latvia show a GDP per capita ranging from 4,900 to 3,600 USD. Bulgaria and Romania, planning accession for 2007 at the earliest, show a GDP per capita around 2,000 USD. The Ukraine is currently not considered a candidate state and has a GDP per capita less than 900 USD.

Unemployment is also a fundamental economic indicator. The highest rates of unemployment in 2001 were in the CEE countries. Slovakia, Bulgaria and Poland were worst afflicted, with around 19.4 % to 18.5 % of the workforce without jobs. The lowest rate of unemployment was in Austria with only 3.6 %, followed by Hungary (5.6 %), Slovenia (5.8 %) and Romania (6.6 %). All these four countries show better figures than Germany, at 7.7 %, though it should be noted that in eastern Germany (the former German Democratic Republic) the unemployment rate is about twice as high as in western Germany.

Table 3. Annual growth rates of GDP 1989-2002 at constant prices (%)

	GDP, constant prices, annual percent change													
	1989	1990	1991	1992	1993	1994	1995	1996	1997	1998	1999	2000	2001	2002
Austria	4.2	4.7	3.3	2.3	0.4	2.6	1.6	2.0	1.6	3.9	2.7	3.5	0.7	1.0
Bulgaria	-0.5	-9.1	-10.8	-8.4	-11.6	-3.5	-1.8	-8.0	-5.6	4.0	2.3	5.4	4.1	4.8
Czech Rep.	4.5	-2.4	-11.6	-0.5	0.1	2.2	5.9	4.3	-0.8	-1.0	0.5	3.3	3.1	2.0
Germany	3.6	5.7	5.0	2.2	-1.1	2.3	1.7	0.8	1.4	2.0	2.0	2.9	0.8	0.2
Hungary	0.7	-3.5	-11.9	-3.1	-0.6	2.9	1.5	1.3	4.6	4.9	4.2	5.2	3.8	3.3
Latvia	n/a	n/a	n/a	-32.1	-11.4	2.1	-0.8	3.7	8.4	4.8	2.8	6.8	7.9	6.1
Poland	3.8	-7.2	-7.0	2.0	4.3	5.2	6.8	6.0	6.8	4.8	4.1	4.0	1.0	1.4
Romania	-5.8	-5.6	-12.9	-8.8	1.5	3.9	8.0	3.9	-6.1	-4.8	-1.2	2.1	5.7	4.9
Slovak Rep.	4.5	-0.4	-15.9	-6.7	-3.7	5.2	6.5	5.8	5.6	4.0	1.3	2.2	3.3	4.4
Slovenia	0.8	-7.5	-8.9	-5.5	1.7	5.8	4.9	3.5	4.6	3.8	5.2	4.6	2.9	3.2
Ukraine	n/a	n/a	n/a	-9.7	-14.2	-22.9	-12.2	-10.0	-3.0	-1.9	-0.2	5.9	9.2	4.8

Source: IMF 2003

The high unemployment rates in the CEE countries are a consequence of the economic slump which accompanied the initial period of transition and went on for several years (see Table 3). While Austria shows a more or less typical western-European pattern of moderate *GDP growth* in the period from 1989 to 2002, all CEE countries went through a phase of diminishing GDP. In Poland the economy only shrank in the years 1990 and 1991. The Czech Republic and Slovenia were able to halt their economic decline after three years. In Germany the short but drastic collapse of the eastern German economy lead to a negative growth rate over the whole German economy in 1993.

Table 4. Sectoral structure of economy, life expectancy (2001)

	Structure of production (% of Gross Value Added) 2001[1]			Life expectancy at birth (years) 2001[2]
	Agriculture	Industry	Services	
Austria	2.2	30.4	67.4	78.3
Bulgaria	13.4	28.7	57.9	70.9
Czech R.	4.3	39.4	56.2	75.1
Germany	1.2	30.4	68.4	78.0
Hungary	4.3	31.3	64.4	71.5
Latvia	4.8	24.8	70.3	70.5
Poland	3.8	31.3	65.0	73.6
Romania	14.7	36.6	48.7	70.5
Slovak R.	4.5	31.8	63.8	73.3
Slovenia	3.3	36.1	60.9	75.9
Ukraine	16.9	35.2	47.9	69.2

[1] Source: European Commission, Reports of candidate countries 2003
[2] Source: United Nations Development Program, HDR 2001

The worst slump in economic activity can be seen in Bulgaria, where GDP sank over nine consecutive years. In the Ukraine the economy was also in contraction for at least eight years. The problems of transformation in Bulgaria and the Ukraine were additionally aggravated by the economic crisis in Russia and war in Kosovo.

A comparison of the *sectoral structure* of the countries' economies gives a useful insight into the state of their development and indicates where (further) structural change is desirable (see Table 4). Most striking is the high share of agriculture in the economies of Ukraine (16.9 %), Romania (14.7 %) and Bulgaria (13.4 %). While Romania and the Ukraine are also characterised by a high share of industrial output, Bulgaria has a comparatively healthy services sector. Agriculture is least important to the economies of Germany (1.2 %) and Austria (2.2 %), in contrast to the prominent position of services (68.4 % and 67.4 % respectively), only matched by Latvia (70.3 %) and Poland (65 %). The countries with the greatest reliance on industry are the Czech Republic (39.4 %), Romania (36.6 %), Slovenia (36.1 %) and the Ukraine (35.2 %).

Finally, one last indicator for the quality of life and the general economic situation should be mentioned: *life expectancy* at birth. This ranges from around 78 years in Austria and Germany to 69/70 years in Ukraine, Latvia, Hungary and Bulgaria.

It is important to note that a direct comparison of statistics can sometimes give a false picture. For example, the *shadow economy* must be taken into account when investigating GDP per capita, GDP growth rates and the structure of production. Thus it is estimated that in 1995 the black economy was vast (over 24 %) in the Ukraine, Bulgaria, Hungary and Latvia, with a lower percentage in Romania. In the Slovak Republic, Austria and Germany (both 1993), Czech Republic and Poland, the shadow economy contributed less than 15 % of economic activity (Friedmann 1999). In a similar way it is difficult to compare the officially recorded unemployment rates. Apart from illicit employment, the methods of determining the unemployment rate differ from country to country. Thus a sensible use of the presented data is to show trends and outline the main differences between countries.

References

Armstrong H, Taylor J (2000) Regional Economics and Policy, 3rd edn. Oxford/Malden

Bachtler J, Downes R, Gorzelak G (eds) (2000) Transition, cohesion and regional policy in Central and Eastern Europe. Aldershot Ashgate

Berglöf E, Roland G (1997) The EU as an „outside anchor“ for transition reforms. SITE Working Paper No. 132

Blakely EJ, Bradshaw TK (2002) Planning local economic development. Theory and Practice. 3rd edn. Thousand Oakes London New Delhi

Blotevogel HH (1997) Strukturwandel und Regionalpolitik im Ruhrgebiet. In: Axt H-J (ed) Strukturwandel in Europa: Südosteuropa, Ruhrgebiet und Reform der EU-Strukturpolitik, Südosteuropa Aktuell 26. München, pp 125-140

Butzin B (1987) Zur These eines regionalen Lebenszyklus im Ruhrgebiet. In: Mayr A, Weber P (eds) 100 Jahre Geographie an der Westfälischen Wilhelms-Universität Münster (1885-1985). Paderborn, pp 191-210

Butzin B (1993) Strukturkrise und Strukturwandel in "alten" Industrieregionen, Das Beispiel Ruhrgebiet. Geographie Heute 113/1993: 4-12

Carter, FW, Turnock D (eds) (2001) Environmental problems in Eastern Europe. London New York

Cooke P (ed) (1995) The rise of the rustbelt. London

Crampton RJ (1997) Eastern Europe in the Twentieth Century – and after, 2nd edn. London New York

Dostál P (1998) Democratisation, economic liberalisation, and transformational slump: a cross-sectional analysis of twenty-one postcommunist countries. Environment and Planning C: Government and Policy, vol 16: 281-306

European Commission (2001) The European Union: Still enlarging. Luxembourg

European Commission (2003) Reports of candidate countries 2003, in: www.europa.eu.int/comm/enlargement/report_2003/pdf, download: November 2003

European Commission (2004) Third report on economic and regional cohesion. Luxembourg
Fazekas K, Gorzelak G (1995) Restructuring the labour market in regions dominated by heavy industry in Central and Eastern Europe. In: Centre for co-operation with the economies in transition, OECD (ed) The regional dimension of unemployment in transition countries. Paris, pp 314-344
Förster H (1996) Altindustrieregionen in West- und Osteuropa. Tübinger Geographische Studien, Heft 116: 21-54
Förster H (1999) Entwicklungsprobleme altindustrialisierter Gebiete im Transformationsprozeß. In: Pütz R (ed) (1999) Ostmitteleuropa im Umbruch – Wirtschafts- und sozialgeographische Aspekte der Transformation. Mainzer Kontaktstudium Geographie, vol 5. Mainz, pp 21-35
Förster H (2000) Transformationsforschung: Stand und Perspektiven. Europa Regional – Zeitschrift des Instituts für Länderkunde, Leipzig, vol 8, No 3/4, pp 54-59
Friedman E et al. (1999) Dodging the Grabbing Hand: The Determinants of Unofficial Activity in 69 Countries. The economics of transition 148.
Gorzelak G (1998) Regional development and planning in East Central Europe. In: Keune M (ed) (1998) Regional development and employment policy. Lessons from Central and Eastern Europe, ILO-CEET. Geneva Budapest, pp 62-76
Gorzelak G (2002) Regional Development in Europe: A historical Perspective, Documentation of the powerpoint slides of a lecture given at the Autumn School 2002 of the School for International Regional Development Studies in Glasgow
Gorzelak G, Ehrlich E, Faltan L, Illner M (eds) (2001) Central Europe in transition: Towards EU membership. Warsaw
Grabher G (1993) Wachstumskoalitionen und Verhinderungs-Allianzen. Entwicklungsimpulse und -blockierungen durch regionale Netzwerke, Informationen zur Raumentwicklung, Heft 11: 749-757
Grabher G (1997) Adaptation at the cost of adaptability? Restructuring the Eastern German regional economy. In: Grabher G, Stark, D (eds) (1997) Restructuring networks in post-socialism. Legacies, linkages, and localities. Oxford, pp 107-134
Grabow B, Henckel D, Hollbach-Grömig B (1995) Weiche Standortfaktoren, Schriften des Deutschen Instituts für Urbanistik 89. Stuttgart
Gros D, Steinherr A (1996) Winds of change. Economic transition in Central and Eastern Europe. London New York
Hamilton FEI (1995) Re-evaluating space: Locational change and adjustment in Central and Eastern Europe. Geographische Zeitschrift, vol 83, 67-86
Hamm R (1991) Umstrukturierungsprobleme in den neuen Bundesländern und Erfahrungsmuster altindustrieller Regionen. Raumforschung und Raumordnung, 49. Jahrgang, Heft 2-3: 91-100
Hamm R, Wienert H (1990) Strukturelle Anpassung altindustrieller Regionen im internationalen Vergleich. Schriftenreihe des RWI Essen 48. Berlin
Häußermann H (ed) (1992) Ökonomie und Politik in alten Industrieregionen Europas. Probleme der Stadt- und Regionalentwicklung in Deutschland, Frankreich, Großbritannien und Italien. Basel Boston Berlin
Herrschel T, Forsyth T (2001): Constructing a new understanding of environment under postsocialism. Environment and Planning A, vol 33: 573-587
Hesse JJ (ed) (1988) Regional structural change and industrial policy in international perspective: United States, Great Britain, France, Federal Republic of Germany. Baden-Baden
Hodgkinson A, Nyland C, Pomfret S (2001) The determination of location in New South Wales. Regional Studies, vol 35.1: 39-55

Hudson R (1997) Regional futures: Industrial restructuring, new high volume production concepts and spatial development strategies in the new Europe. Regional Studies, vol 31.5: 467-478

Informal meeting of the ministers for spatial development (ed): European Spatial Development Perspective, Glasgow 8th June 1998

IMF – International Monetary Fund (2003) World Economic and Financial Surveys, in: www.imf.org/external/pubs/ft/weo/2003/02/data, download: November 2003.

Keune M (ed) (1998) Regional development and employment policy. Lessons from Central and Eastern Europe, ILO-CEET, Geneva Budapest

Kloten N (1993) The time dimension of the transition of economic systems. In: Wagener H-J (1993) On the theory and policy of systemic change. Heidelberg New York, pp 121-133

Kunze C (ed) (1997) Strukturkrisen altindustrialisierter Regionen und ihre Bewältigung in West- und Osteuropa. Transformation – Leipziger Beiträge zu Wirtschaft und Gesellschaft, vol 6. Leipzig

Kwiatkowski E, Janusz T, Steiner V (1995) Unemployment in light industry regions: An overview of Eastern Europe and a case study of Lodz Region in Poland. In: Centre for co-operation with the economies in transition, OECD (ed) The regional dimension of unemployment in transition countries. Paris, pp 269-313

Lintz G, Müller B, Schmude K (2004) The future of industrialised cities and regions in Central and Eastern Europe, completely revised version of a paper given at the conference "The end of socialism...? 10 years on" of the Royal Geographical Society, Post Socialist Research Group, London, 4th-5th May, 2001, submitted for publication to Geoforum

Mayr A, Grimm F-D, Buchholz HJ (eds) (2001) Transformationsforschung: Stand und Perspektiven. Beiträge einer wissenschaftlichen Tagung. Europa Regional, Special Issue of the Zeitschrift des Instituts für Länderkunde, Leipzig, vol 8, No 3/4

Molle W, Verkennis A (1995) Regional development and changing location factors: a challenge for regional and local authorities. EUREG 2/95: 42-47

Murray R (1992) Flexible specialisation and development strategy: The relevance for Eastern Europe, in: Ernste H, Meier V (eds) Regional development and contemporary industrial response. Extending flexible specialisation. London New York, pp 197-218

Myrdal G (1957) Economic theory and underdeveloped regions. London

Pavlínek P, Pickles J (2000) Environmental transitions. Transformation and ecological defence in Central and Eastern Europe. London New York

Pickles J, Smith A (eds) (1998) Theorising Transition. The political economy of post-communist transformations. London New York

Pindyck RS, Rubinfeld DL (1998) Microeconomics, 4th international edn. New Jersey

Pütz R (ed) (1999) Ostmitteleuropa im Umbruch. Wirtschafts- und sozialgeographische Aspekte der Transformation, Mainzer Kontaktstudium Geographie 5. Mainz

Roland G (2001) Ten years after... Transition and economics. In: International Monetary Fund (ed) Transition economies: How much progress? IMF Staff Papers, vol 48, special issue, pp 29-52

Sadler D (2004) Cluster evolution, the transformation of old industrial regions and the steel industry supply chain in North East England. Regional Studies, vol 38.1: 55-66

Sajdik M (2004) Willkommen in der Union: Finanzhilfen für den Beitritt. RAUM, 53/04: 10-13

Samuelson PA, Nordhaus WD (1998) Economics, 16th international edn. Boston

Schrader M (1993) Altindustrieregionen der EG. In: Schätz L (ed) Wirtschaftsgeographie der Europäischen Gemeinschaft. Paderborn, pp 111-166

Smith A (1996) From convergence to fragmentation: uneven regional development, industrial restructuring, and the 'transition to capitalism' in Slovakia. Environment and Planning A, vol 28: 135-156

Stadelbauer J (2000) Räumliche Transformationsprozesse und Aufgaben geographischer Transformationsforschung. Europa Regional 8(2000)3/4: 60-71

Steiner M (1985) Old industrial areas: A theoretical Approach. Urban Studies 22: 387-398

Tichy G (1987) Das Altern von Industrieregionen. Unabwendbares Schicksal oder Herausforderung für die Wirtschaftspolitik? Berichte zur Raumforschung und Raumplanung, 31(1987)1/2: 3-10

Tickle A, Welsh I (eds) (1998) Environment and Society in Eastern Europe. Harlow New York

United Nations Development Programme (2003): Human Development Report 2001, in: http://hdr.undp.org/reports/global/2001/en/pdf/completenew.pdf, download: November 2003

United Nations Economic Division for Europe (2003) Statistics 2003, in: www.unece.org/stats/trend/ukr.pdf, download: November 2003

Vanhove, N (1999) Regional policy: A European approach, 3rd edn. Aldershot 1999

Werner K (1995) Die Lage der DDR-Wirtschaft am Vorabend der Währungs-, Wirtschafts- und Sozialunion. In: Pohl R (ed) Herausforderung Ostdeutschland. Fünf Jahre Währungs-, Wirtschafts- und Sozialunion. Berlin, pp 35-51

Wood G (1994) Die Umstrukturierung Nordost-Englands. Wirtschaftlicher Wandel, Alltag und Politik in einer Altindustrieregion. Duisburger Geographische Arbeiten 13. Dortmund

Wood G (1997) Social and economic restructuring in old industrial areas: A comparison of North East England and the Ruhr district in Germany. In: Blotevogel HH, Fielding AJ (ed) People, jobs and mobility in the new Europe. Chichester, pp 189-215

Wößmann L (2001) Der Aufstieg und Niedergang von Regionen: Die dynamische Markttheorie von Heuß in räumlicher Sicht. Jahrbuch für Regionalwissenschaft, vol 21: 65-89

Websites:

EU website for enlargement: http://europa.eu.int/comm/enlargement/index_en.html, Mai 2004

EU website for external relations : http://europa.eu.int/comm/external_relations/ceeca/index.htm

EU website for regional policy: http://europa.eu.int/comm/regional_policy/index_en.htm, Mai 2004

Website of Euractive: http://www.euractive.com

Austria: The Process of Restructuring is Largely Complete

Gerhard Schimak, Thomas Dillinger and Helga Wagner

Vienna University of Technology, Institute of Urban Design and Planning, Vienna

1 Introduction

The following contribution outlines the role of industry in Austria's historical development, the structural problems caused by economic and geopolitical change as well as the economic and industrial-political measures for overcoming these problems. To begin with it is necessary to point out the specific Austrian situation, within the context of traditionally industrialised cities and regions. As opposed to the countries of Central and Eastern Europe (CEE States) and Germany, very few mono-structured "industrial regions" were formed as Austria underwent industrialisation. And those that were formed achieved relatively low expansion by international comparison. It is therefore futile to search for large, mono-structural industrial regions within small-structured Austria. In this context it should be clear that typical problems of structural crisis in industry, such as sweeping sectoral job cuts leading to massive national unemployment, had a less dramatic impact in Austria than in Eastern European countries and affected only small select regions.

As many of the activities which helped shape the current situation are rooted in the country's past, a preliminary look at historical development is essential. The development of Austrian industry has been largely determined by geography (Alpine region), political events (the fall of the Habsburg Empire, World War II and occupation) and state industrial policy (state-controlled industry). Under the Second Republic, nationalised industry played a traditionally important role in the Austrian economy (especially in the 1960s and 1970s). This matter will therefore be afforded particular attention. Indeed it has been the ongoing decline of state industry since the 1970s that has laid the foundation for the problems which many traditionally industrial European countries currently face.

Today the process of restructuring is largely complete. The various regions and sectors have confronted various difficulties and solved them in different ways. The causes and effects of specific problems as well as the means and strategies used in the restructuring process will therefore be discussed. Furthermore, we will describe those models and measures on each decision-making level (state government, regional government and, since 1995, the European Union) which have an influence on the general development of industry in (old) industrial regions.

2 The Situation of Industrial Regions and Cities in Austria

2.1 Historical Development

In pre-industrial times, economic activity in Austria was centred on the abundant Alpine mineral resources (gold, silver, lead, copper, salt, and iron). Gold, silver and lead mining, known in the south-western regions of Salzburg since the times of the Roman Empire, experienced an upswing in the 15th and 16th centuries. However increased competition from South American silver mines, overexploitation, the exhaustion of pits and a scarcity of wood led to a loss of importance for gold and silver mining. These factors also later contributed to the decline of other industries which exploited mineral deposits (e.g. iron and steel industry).

Salt mining was of fundamental importance during the late Bronze and Iron Ages (ca. 740-450 BC). It was practised all along the northern Alpine border, among other places in Hallstatt. In the early 16th century production was monopolised by the state.

The extraction of iron ore, which can be traced back to the 7th century BC, attained greater significance under the Celts. The most important deposits were concentrated in the Styrian Erzberg, ranking among the most profitable mining regions in Europe during the Middle Ages. In the 14th century, with the new technology of iron-processing in hammer mills, areas rich in forest and water became important in iron mining and processing. A European centre for the small-scale iron industry formed in the so-called “Eisenwurzen” region north of the Erzberg mountain. The iron-processing businesses were connected to the Erzberg by a diversified transport system (rafts on water, the so-called “Iron Road” on land). As a cultural and economic unit, Eisenwurzen and Erzberg became a region based on the division of labour. In the 16th century the country’s ruler introduced a system of designation (“Widmung”) to the iron industry in order to optimise production. Under this regulation, certain agricultural areas and forests were designated exclusively for iron industry production. The strict system of designation was an important factor for the success of iron production, stabilising regional structures and securing important conditions for the production process. The eastern regions of the Austro-Hungarian Empire received special attention from the state government during the period of industrialisation. In particular, the industrial areas in Bohemia, Moravia and Silesia were developed at that time. However with the collapse of the monarchy in 1918 Austria lost these important industrial regions, along with rich hard coal deposits in Upper Silesia. Subsequently, the mining of (less valuable) brown coal replaced that of hard coal in the small Austrian state, especially in the regions of northern Styria and southern Lower Austria.

The industrial areas still remaining in Austria are concentrated in the southern part of the Danube river basin, the Styrian Mur-Mürz-Furche region and in the Rhein Valley to the west (see Map). Based on its present national boundaries, Austria can be said to have been an "industrialised agrarian state" during the late 19th and early 20th centuries (Bamberger et al. 1995, 548).

Protectionism in the steel and iron industries, as practised by the successor states to the Austro-Hungarian Empire, destroyed Austrian export markets in this sector. During the world economic crisis from 1929 to 1938, the Austrian iron industry suffered a complete collapse, leading to high unemployment. At the same time, the spectrum of Austrian industrial branches was able to expand due to the development of oil fields during the inter-war period (1918-1938). The first successful drilling took place in 1934 (Lichtenberger 1997, 132).

With the annexation of Austria by the German Third Reich in 1938, Austrian industry again experienced an upswing. In the iron and steel sector, for example, the auctioning of ore mining in the Styrian Erzberg triggered a massive economic boom in the region. Even though the German arms industry brought a period of economic prosperity to the various industrial regions, the boom remained one-sided and tailored to the demands of war. The focus was on energy, raw materials and heavy industry.

After World War II, large sections of Austrian industry were proclaimed "German property" by the Potsdam Agreement of 1945 and turned over to Allied occupational forces as reparations for damages suffered during the war. The following ten years of occupation would significantly determine the dynamics of economic development in the various industrial regions of Austria. While regions under the control of western occupational forces (western and southern regions) could develop relatively freely (see the example of Vorarlberg in section 3), former "German property" in the eastern industrial regions under Soviet control were absorbed into the USIA-Group[1] and began producing for the "Eastern Bloc" states. To prevent the loss of these important businesses and their economic potential, the Austrian government introduced a state control programme in 1946, under which an initial 70 businesses largely involved in heavy industry (coal, iron ore, petroleum, metal, electricity, machine and chemical industries) were nationalised (with the exception of the USIA-Group) (Lichtenberger 1997, 274). The Marshall Plan helped both nationalised and private-sector businesses to rebuild and modernise. State industry soon became the driving force behind the reconstruction of the Austrian economy, producing all raw materials and the majority of semi-finished products. With prices far below those on the world market, subsidies were given to resettled industrial areas, a measure from which the machine industry would particularly stand to benefit. The strong position of state industry is reflected in the following figures: between 1945 and 1970 the state sector was responsible for over one fifth of the gross national product, and oversaw the doubling of productivity per employee and the tripling of production value. A third of Austrian exports could also be attributed to state industrial businesses (Lichtenberger 1997, 275).

[1] USIA: Administration of Soviet Property in Austria in the post-war period.

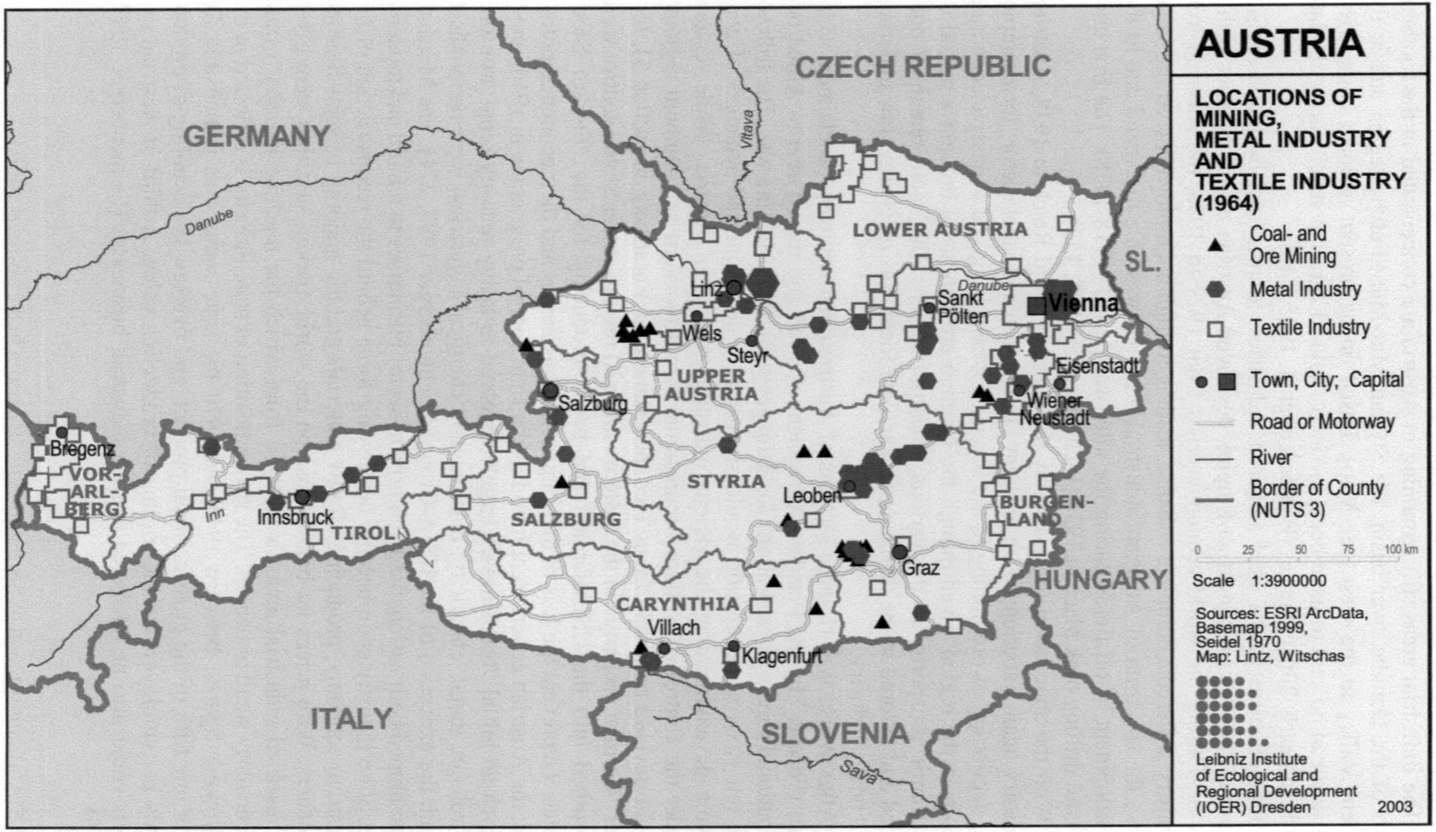

Map: Austria – Distribution of mining, metal industry and textile industry in 1964

A map from 1970 (see above) shows core regions for the most important industries at that time: textiles, metallurgy, coal and iron ore. The textile industry is concentrated in a number of small businesses in western Vorarlberg, in the Danube basin south of Vienna and in the northern border region Waldviertel (Upper Austria). Metallurgy mainly exists in the form of small and medium-sized businesses along the Styrian Mur-Mürz-Furche. And coal and iron mining industry can be found selectively in northern Styria (Erzberg/iron ore) and southern Austria (brown coal).

In the post-war period the price of raw materials fell dramatically on the international market. Because Austrian companies were unable to compete, many businesses in the historically significant sector of mountain mining (salt, iron, precious metals) were forced to close. State industry took on an active role in regional and labour politics in the fight against unemployment, with state regulations helping to maintain existing structures for some time. However, weakened by high long-term losses, the industrial group finally began the necessary process of restructuring in 1986, with the passing of the so-called ÖIAG[2] law. This law supports the reorganisation, renovation, alienation and liquidation of deficient group sections. The restructuring of export-oriented state industry was largely complete by the late 1980s and early 1990s (prior to EU accession) and only a small number of businesses still required restructuring (Soukup 1996). Because all important industries were encompassed by state industry, the history of the state group had a strong influence on the development of typical Austrian industrial regions.

Today, approximately one third of the nominal capital of industrial companies organised in capital investment firms is still state-owned. About 20 % of industrial employees work in these businesses. Yet the fact that extensive privatisation programmes are under way indicates that the percentage of publicly-owned nominal capital will presumably decrease substantially. In any event, Austrian industrial structure is characterised by a large number of small and medium-sized companies, with very few large firms. In general the trend from an industrial to a service-oriented society is spreading into traditional industrial regions. The Eisenwurzen region, for example, has already begun to focus on soft tourism and marketing of its regional "industrial" features. The Styrian Erzberg region, where mining activities will soon cease, began a process of reorientation by supporting the use of opencast pits as unusual venues for diverse cultural events.

2.2 Regional Distribution of Industry

During the 1970s and 1980s an east-west divide was still discernible in Austria. While industry was developing more dynamically in western regions, development in the east was stalled until 1955 and after as a result of the Soviet occupation. The peripheral market position of the eastern border regions along the "Iron

[2] ÖIAG: Österreichische Industrieholding AG (Austrian industrial shareholding).

Curtain" also proved disadvantageous (Hammerer/Grundler 1995, 44-45 and 52-60, Faulhaber 1992, 66-69).

Large-scale industrial regions in Central and Eastern Europe such as the German Ruhrgebiet do not exist in Austria today. Table 1 shows the low rate of industrial employees per 1000 inhabitants, as well as the distribution of industrial businesses and employees (regions correspond to the European regional level NUTS 2). The individual industrial locations are spread in varying spatial and sectoral concentrations across the entire Austrian territory (small by international standards). An overview of the importance of industry in Austrian regions will be presented on a regional level in Table 2.

Table 1. Number of industrial businesses and employees (1994), Industrial employees (1994) per 1000 inhabitants (1991)

Region	Industrial businesses (number)	Industrial employees (number)	Ind. empl. per 1000 inhabitants
Burgenland	210	10,175	38
Lower Austria	1,452	85,886	58
Vienna	1,206	88,556	58
Upper Austria	1,502	112,597	84
Styria	1,158	77,051	65
Carynthia	555	27,412	50
Salzburg	517	23,955	50
Tyrol	620	29,317	46
Vorarlberg	488	28,185	85

Source: ÖSTAT, 1996

With respect to the structure of the different branches, Austrian industry is still centred on primary industry. The following section outlines the situation in the most important industrial branches. The *steel sector* – including state-owned companies – is still the most important sector in Austrian industry. Statistically grouped with the metal and machine industries, the steel branch was responsible for 20.7 % of Austrian gross domestic product (GDP) in 1994, employing a fourth of the industrial workforce. The steel industry held a leading position, both in terms of economic performance and the job market. The *vehicle, electric and metalware branches* developed very dynamically. These branches conduct intensive research, maintain high technical standards, produce high added value and employ highly qualified staff.

The second largest industrial sector, the *chemicals industry*, has serious structural problems. In low-tech areas (paint, enamel, plastic and detergents), exports have been declining for years. In the innovative industries (pharmaceutical, gene technology) Austria is largely export-independent. A sweeping restructuring process should soon take place in the field of low-tech production.

Table 2. Overview of the regional distribution of industrial businesses in Austria

	Industrial significance	Most important industrial branch	Other important industrial branches
Burgenland	low, due to late industrialisation (1930s)	food and tobacco, clothing	
Lower Austria	great importance, dynamic development around Vienna, problem areas in northern border zones	iron and metal ware, chemicals, machine and steel building	petroleum (majority of Austrian oil fields, only refinery), food industry
Vienna	special case due to high settlement, strong industrial location, businesses mainly at city periphery	machine and steel building, vehicle industry, electronics industry, chemicals industry	
Upper Austria	great importance, esp. city of Linz and southwestern surroundings, strongly affected by crises in state industry	machine and steel building, chemicals industry, iron production	vehicle industry, iron and metal ware
Styria	great importance, traditional iron industry (Mur-Mürzfurche), two important mining locations currently in decline	mining and iron production (about 50% of the national production value)	paper, leather industry, machine and steel building, iron and metal ware, saw industry, electricity and electronics industry, vehicle building, cement industry
Carynthia	low importance	metal processing, stone and ceramic industries, chemicals industry	food and tobacco, wood processing
Salzburg	lowest industrial significance in Austria	machine and steel building, food and tobacco, chemicals industry, wood processing, iron and metal ware	
Tyrol	low importance, mountains limit site location, sites esp. in the Inn Valley	glass production	metal branch, stone and ceramic industries, leather production, clothing and textile sectors, wood processing, chemicals industry
Vorarlberg	(old-) industrial region, countless industrial sites in the Rhine Valley, very dynamic industrial market	textile and clothing sectors (1/3 of regional industrial production)	iron and metal ware, food and tobacco, machine and steel building, electricity industry

Source: Faulhaber 1995, 65-69; new presentation

The *textile and clothing industry* experienced a serious decline after several crises in the post-war era and in the 1970s. While this sector showed the largest percentage of employees in 1961 at 17 %, by 1994 only 8.1 % of industrial employees were working in the textile and clothing sector. While most textile-dominated regions such as the Waldviertel were not able to overcome these structural crises, Vorarlberg managed to reorient itself on the market through specialisation, technological innovations and improved quality. Today the Vorarlberg region is a perfect example for the successful restructuring of a main industrial branch.

The extraction of *mineral deposits* is reduced to a minimum today. Gold and silver mining have been entirely abandoned, and salt mining is only practised to a small extent. Iron ore and coal mining have also been temporarily halted, whereas Austrian magnetite mining has retained its importance and achieved a high international standard.

3 Special Problems of Industrial Cities and Regions

Arising from the country's historical development, the specific problems of Austria's declining industrial cities and regions are clearly delimited. Here it is important to realise that Austria never experienced industrialisation on a large scale. Instead, industrial businesses were located in small regions and isolated sites. The structural shift in Austrian industry is also largely complete, and industrially dominated problem regions are already moving on to new economic perspectives (such as in the Eisenwurzen region).

In general, problems connected with the industrial situation were based on two factors. The first was the regional *industrial monostructure* (not restricted to Austrian industrial regions), in turn embracing many other problems:

- high unemployment, due to sectoral job loss and difficulty in reorientation on the job market,
- few alternatives – the lengthy dominance of large industrial businesses had blocked the rise of new companies,
- a lack of entrepreneurial spirit,
- social tensions among employees, resulting from the loss of company-based social activities and job-related prestige, etc.

Other problems in Austria were directly related to the country's *topography*. The lack of usable land for business expansion greatly inhibited economic growth among industries located in the Alps, leading to a complicated transport and delivery system that adversely affects competitiveness. In addition to such general aspects, branch-specific problems can be identified. These will be examined more closely by focusing on three important industrial branches: the textile, iron and chemical branches.

Textile industry: As can be seen on the map, in the 1960s the Austrian textile industry was still mainly concentrated in the Danube basin, in northern Lower Austria (Waldviertel region) and in Vorarlberg. Since the 1970s intense national competition has forced the relocation of production to developing countries and countries of central eastern Europe, resulting in the closure and restructuring of businesses. Large companies in the Danube basin or the peripheral Waldviertel did not react quickly to counter this global trend: their inflexible production organisation thwarted necessary and rapid adaptation, whereas in Vorarlberg, where many textile producers still kept farms, the large number of smaller businesses or "home factories" were able to work more flexibly and remain competitive. In times of crisis this structure acted as a buffer, helping a large number of the unemployed to keep their heads above water by means of farming. Such small-scale structures also left room for the development of other industrial branches.

Jobs lost in Vorarlberg were, to an extent, absorbed by other industrial branches (metal, machine and electricity industry). After World War II, Vorarlberg's situation in the Allied-occupied zone proved to be highly advantageous. Whilst industrial facilities in Soviet-occupied eastern Austria were being torn down as reparations for war, Vorarlberg already had access to the western trading area. With the creation of the "Economic Agency Vorarlberg-Switzerland" in 1945, the foundation stone was laid for the later economic boom in the Vorarlberg textile industry (Gebhardt 1990, 98f.). A later crises in the textile branch may have led to a decline in the number of businesses, but many companies in Vorarlberg were able to re-established themselves by adapting their products and procedures (shifting to high-fashion and niche products).

Iron producing and processing industry: This branch also experienced many problems due to inflexibility (e.g. state "designation" under the Austro-Hungarian Empire) and the size of organisational structures (esp. state industry). The Eisenwurzen region, after its heyday under the Austro-Hungarian Empire, could neither undertake the necessary structural adaptation (because of state "designation") nor withstand the pressures of new technologies and world economic change. Iron industry on the Erzberg itself has managed to survive until today, but even this bastion is posed to fall. For the region, this means that existing problems such as unemployment and difficulties in economic reorientation will be exacerbated.

Chemicals industry: This important Austrian industrial segment has undergone deep restructuring in low-tech production. This applies to the production of automobile tires (Semperit factory in Traiskirchen), cellulose (Lenzing), and plastic household items, where high investments in environmental protection facilities are undermined by a continual price drop. A further problem is that the majority of medium-sized businesses tend to function as dependent subsidiary branches of multinational corporations. This lack of regional anchoring encourages a tendency toward production relocation abroad.

4 Reasons for the Decline of Industrial Regions

An analysis of the reasons for the decline of Austrian industrial cities and regions clearly exposes the influence of other economies on domestic industries. Global developments are weakening industrial cities and regions, for instance through production relocations to lower wage countries (currently felt in the Austrian textile industry). Monostructuring, which makes regions more susceptible to the disturbance or elimination of their dominant economic branch, can also promote decline and lead to other problems. These include, for example, a lack of educational facilities for vocational retraining, low site attraction for new companies, and a need for economic reorientation.

Austria has experienced two particularly serious crises in the industrial sector. Both the early industrial boom during the Austro-Hungarian Empire and the later economic upswing after World War II (following nationalisation of most industrial branches) were followed by serious structural crashes that initiated a decline in prospering industries. Surprisingly, precisely those structures which had ensured the success of industrial regions became elements of weakness under the new circumstances. Although some of these developments can be considered historical, they are still relevant examples of the reasons for regional industrial decline.

Economic life under the Austro-Hungarian Empire was shaped by a political leadership that permitted very little economic freedom. The enormous internal market was controlled by a protective tariff policy, promoting the formation of cartels or similar agreements among entrepreneurs (Gebhart 1990, 62; Faulhaber 1992, 17-18). Strict state regulations cemented traditional structures and stalled the necessary process of adaptation required in a changing global market (especially in the steel sector) and modern production methods. An additional obstacle was the lack of available land for expansion in the Alpine region. The collapse of the Austro-Hungarian Empire and its economic area after World War I created new basic conditions: the small and almost exclusively alpine state

Austria was formed from the remnants of the empire. The remaining industry had about one third of the original capacity, though even this was oversized for the reduced domestic market. Furthermore, owing to the long tradition of protective tariffs industrial branches were underdeveloped (Gebhart 1990, 62). For some branches the highly protectionist economic policy in neighbouring countries also meant the loss of important industrial materials and markets. For many Austrian industrial regions such as the metal-producing Eisenwurzen region, this development meant the loss of their main source of income.

After World War II, companies that had been nationalised were the motors for reconstruction in the small Austrian state (see also section 2). Since then state industry has played an important role in the Austrian economy, and nearly all key and heavy industries (metal, iron, petroleum, coal, electricity, machine and chemical industries) are state-run. State industry is a conglomerate operating in over 400 fields of businesses. For this reason, the development or decline of many industrial regions in Austria is closely linked to the fortunes of nationalised industry and their individual companies.

During the world economic crisis in the early 1970s, state industry assumed an important role in labour and regional politics (Gassner and Simonitsch 1987, 103). The proclaimed goal was to achieve full employment, while accepting lower productivity and economic losses. However, state regulation resulted in the delay of necessary adaptation procedures. State-run companies were never quite able to escape political influence, and the composition of all management levels in companies corresponded exactly to the distribution of power among the government coalition parties. "Party-line economics" and proportional representation (Proporz) was sometimes more important than professional qualifications. Political influence on management prevented an economically sensible approach to administration. The situation finally proved unacceptable under the pressure of enormous debts, and in 1986 the so-called ÖIAG law[3] was passed. This was a necessary step toward the restructuring of the entire group; the law aims at the reorganisation, renovation, alienation or liquidation of deficient group companies (Lichtenberger 1997, 276). State industry then began the process of privatisation, a process which is bound to continue and will lead to industry job losses.

5 Strategies for the Development of (Old) Industrial Cities and Regions

While in the 1960s and 1970s strategies for industrial regions and cities focused on attracting companies and retaining or creating jobs ("quantitative" industrialisation policy, growth oriented), economic policy in the 1990s focused on endogenous regional and innovation policy. The term innovation is here not limited to research and development, but also includes political/administrative structures and processes (http://www.industriellenvereinigung.at). Support is given to large companies, as in the past, with additional measures to subsidise SMEs (small and medium-sized enterprises), in keeping with EU policy.

State, regional and European government all play a decisive role in Austrian regional and innovations policy. The federal structure of Austria means that the regions are of particular importance. In joining the European Union in 1995, the country gained access to industrial subsidies and spatial-structural programmes (EFRE) that support new enterprises.

5.1 Strategies in EU Regional Policy

After accession to the EU, Austria's economic and industrial policies became closely linked to EU goals. In regard to (old) industrial regions, several "Objective 2" regions were identified (in the previous programme period 1994-1999, these were designated "industrial regions with recessive development"). They include parts of Vorarlberg, Lower and Upper Austria and half of Styria. Former textile regions such as the Waldviertel, southern Lower Austria, eastern Styria and parts of Vorarlberg also received subsidies through the community initiative RETEX.

---

[3] ÖIAG: Österreichische Industrieholding AG (Austrian Industrial Shareholding) consisting of VOEST-Alpine Stahl AG, M&A Holding (machine and facility building), Austria Metall SG (AMAG, Aluminium), OMV (petroleum), ÖBAG (mining), SEH (environmental technologies), Chemie Holding.

The RECHAR II programme, which promotes the economic restructuring of coal-mining areas, was mainly applied in southern Styria. RESIDER II, an initiative that subsidised the economic restructuring of steel areas, was implemented in the regions of northern Styria. According to the Agenda 2000, the total area eligible under Objective 2 in Austria was re-defined and reduced. Traditional industrial regions such as Dornbirn in Vorarlberg, Steyr/Kirchdorf in Upper Austria and parts of southern Lower Austria were no longer considered eligible for subsidies. The RETEX, RECHAR II and RESIDER II programmes were stopped.

5.2 State Strategies

Innovation policy: Since 1990 the Federal Ministry of Trade, Commerce, and Labour[4] has implemented the programmes "Regional Infrastructure Subventions" (RIF) and "Regional Innovation Prize" (RIP) in order to achieve economic renewal in traditional industrial regions and vitalise peripheral regions. They are both financed by the federal and regional government. RIF supports the creation of impulse centres to promote regional growth, including the founding of technology (transfer) centres. RIP helps individual companies with the aim of forcing a change in the branch structure, rejuvenating the regional economic structure and creating and attracting companies (Birner et al. 1999).

Labour market policy: In addition to the innovation policy measures already mentioned, the federal government has also taken steps to promote the labour market. According to §§ 27a, 35a and 51a of the Labour Market Promotion Law (AMFG, BGBl. Nr. 313/1994), the labour market service offices are required to practice an active labour market policy by offering company-based subsidies for restructuring and investment aimed at securing and creating jobs. Small and medium-sized companies, companies in problem zones with high structural unemployment and low economic performance and companies with special importance on the labour market and in regional politics are given preference.

Local labour market service offices in areas with recessive industrial development have in part carried out special labour market projects, placing special emphasis on the temporary institution – "job foundation" – as a public-private partnership. It was assumed that employees who had been made redundant from large state companies after many years employment would necessarily lose social contacts and side benefits connected with company services and facilities. However the foundation has made it possible to maintain ties between the newly unemployed and their previous employers. The main difference between the labour foundation and typical educational programmes is that schooling is given to those seeking work while they still enjoy the benefits of their previous employment. Information is also continually available about job opportunities arising in the former company. This mitigates the transition to the external job market, while ties are maintained with the former company and its social network. Financing comes from a variety of sources. In the case of the "steel foundation VOEST

[4] Until 2000 the Federal Ministry of Commerce and Traffic.

Linz" the programme was funded by the labour market service, participants in the further education programme, company employees (solidarity fund) and VOEST Linz.

Endogenous regional growth: With the view that winning "external" investors for structurally weak industrial regions can make companies too dependent on multinational corporations, endogenous development strategies have gained importance in Austria in the past decade. This approach follows a policy of transferring responsibility and options for action to the affected regions themselves. By making use of regionally available resources, and by promoting regional identity and activating regional responsibility, this policy of regional anchoring should promote the sustainable development of stimuli within the region itself.

In 1980 the F.E.R. project (a support project for independent regional development) was created by the federal government for the financial support of endogenous regional growth. The main goals were to stimulate endogenous capacities, support niche products and specialised industry and to promote regional co-operation structures among companies.

In the initial years the primary aim was to support companies in peripheral regions. Unfortunately a lack of consulting procedures led to inefficiencies. The plan was then revised to include both consulting and loans (up to 30 % of the total sum). The Austrian Working Group for Independent Regional Development (ÖAR) was founded and now provides the necessary consulting services. In recent years, the development of regional concepts and the introduction of qualified staff in regional management have also been promoted. The F.E.R. is mainly implemented in subsidised areas designated by the Austrian Development Conference (ÖROK 1999, 96 f.).

5.3 Regional Strategies

Economic programmes, goals, and measures: The importance of industrial, economic, technological, and innovation policy in regional government varies according to the degree of industrial orientation. In Styria, Upper Austria, Lower Austria and Vorarlberg the important role played by these policy factors is reflected in their economic programmes. The following goals apply to all regional programmes (http://www.tmg.or.at/ StrategischesProgramm/3.htm, http://www.industriellenvereinigung.at):

- improvement in knowledge base and information transfer (subsidising of future technologies; technology and knowledge transfer; co-operation with business, science and management; development of human capital);
- business foundation offensive;
- development of infrastructure (esp. impulse centres) and increased external orientation (export offensives and promotion of international co-operation);
- image improvement (support of "soft location factors");
- reorientation of state preconditions (administration reform, budget policy).

The general policy of support among traditional industrial regions has particularly centred on structural change, technological and economic modernisation and specialisation on niche products. The strategies mentioned above can be found in the Austrian Objective 2 programme plans for 2000-2006; these have been specially tailored to each region (NUTS 2) despite similarities between goals and measures (http://www.inforegio.cec.eu.int).

The strategy of Lower Austria: Lower Austria has concentrated on soft strategies to realise the endogenous growth potential. The goal is to secure the long-term health and growth of economic locations and an improved regional economic structure. EU-funding and national co-financing mainly support regional management, which in turn supports regional consulting and network-building. Furthermore, this strategy supports the creation of small-scale regional development concepts, planned in collaboration with actors in the region. Support is also given to individual companies. The EcoPlus development agency was founded to assume tasks such as site consulting and company co-operation support. Various other programmes in the region promote soft industrial trade measures, market research, ecological consulting and consulting in the fields of economy, innovation and young entrepreneurs. Regional innovation centres (RIZ) were also set up, in co-operation with the federal government, to promote industrial research and development.

The strategy of Styria: Economic policy in Styria is highly growth-oriented and defines internationalisation as a significant goal. The formation of clusters (metal/tools, wood/paper, transport vehicles/transport) and product group specialisation are regional policy aims and the support of regional economic strongholds in international competition therefore plays an important role. These efforts are underlined by an emphasis on:

- the creation of impulse centres;
- the founding of technology transfer facilities which support research and innovation in business, provide consulting services and help companies to form networks;
- the support of young companies (especially SMEs) by training management and staff, facilitating start-ups and preparing businesses for the information society.

Styria has an especially strong potential in the field of research (two universities, the largest number of patents). An above-average number of impulse centres were set up near research facilities, coinciding with the traditional industrial areas. Advantageous location conditions and strong leading companies contributed to the recent creation of an automobile cluster. In addition to the previously mentioned goals and measures, regional endogenous growth has also been promoted. To this end the regional government of Styria has created the STEFREI project (Styrian Support Project for Independent Regional Initiatives) to promote the development and maintenance of internal regional economic cycles.

From Industry to Tourism Through Regional Management – The Eisenwurzen Region –

The Eisenwurzen region in lower Austria serves as a splendid example of a former wealthy and densely populated industrialised region which, following a history of serious crises and decline, has been able to find a new identity and invigorate its economical development.

Despite continuing problems of low regional productivity, unemployment and the migration of its young people, the region is establishing a new economic orientation by strengthening soft tourism. Several approaches are aimed at supporting the development of the former iron and mining region using the available resources of an industrial heritage and a unique landscape shaped by industry. The region's endogenous regional development policy focuses on the potential of human resources and tries to motivate people, with their specific knowledge and abilities, to participate in the planning process. One association of municipalities aimed to conserve regional historic mining buildings and complexes, while at the same time establishing museums, exhibitions, hiking trails, special events etc. Other projects were the development of a regional innovation centre with integrated technology park, producing a catalogue for regional high-tech products and creating a regional bourse for small-business premises and areas. Most decisive for the success of the Eisenwurzen Region has been good regional management.

Today's institutions of regional management in Austria were mostly created in the early 1990s; the goal was to foster regional development with a special instrument that supports the region's position within the European structural policy, as well as realising more common development tasks. Regional management was created as a multi-functional interface, turntable and networking element, which informs and networks local actors and cooperates with federal, national and EU institutions. Regional management has four fundamental tasks: informing, consulting, project development and development of the regional network of participants. The information aspect concentrates on the main regional topics and key projects. Primary consulting is activating, interlacing and process-accompanying support. Project development in terms of activating and interlacing participants, supervision and leadership of projects or project sponsorship has become the key central function of regional management. Regarding the development of the regional network of participants, one crucial point for network quality is the owner relationship with the municipalities.

Thomas Dillinger and Helga Wagner, Vienna University of Technology, Institute of Urban Design and Planning, Vienna

(References for this case study are given under the general references for the Austrian country report.)

The strategy of Vorarlberg: With its economic and political principles and strategies, Vorarlberg has particularly addressed the problems of the textile industry by supporting diversification within traditional industrial sectors (http://www.vlr.gv.at/). Vorarlberg's textile industry was therefore able to re-establish itself during a severe crisis in the early 1990s by securing niche markets. As well as supporting the textile industry, emphasis was placed on research and development activities and endogenous regional growth.

Organisation structures: In Austria the designation of industrial estates and the regeneration of derelict business locations is the responsibility of municipalities. The regional development laws[5] do not require the elaboration of committed regional programmes for the co-ordination of business areas or the formation of committees for co-ordination purposes. This in part leads to counterproductive, competitive behaviour among communities. To counteract this, organisations have been set up by the individual regions to provide co-ordination and consultation, and to aid the development of economic and technology parks.

With the exception of Vorarlberg, all Austrian regions decided to form societies under direct or partial regional government ownership to help carry out economic goals. The authorities in Vorarlberg chose to leave the development of economic parks to the private economy, with the justification of projected savings and better flexibility. Other benefits are a lower degree of political influence and improvements in the external image or the region, as many entrepreneurs have reservations about "local government agencies". Examples of special organisations for regional development are:

- Lower Austria: ECO-Plus (site, regional, economic consulting), RIZ'-Holding (runs impulse centres);
- Styria: Styrian Association for Economic Promotion (SFG), Innofinanz, Styrian Shareholding;
- Upper Austria: Upper Austria Technology and Marketing Company for Regional Site and Innovation Policy (TGM) (regional site and innovation policy);
- Vorarlberg: PRISMA GmbH (site, regional, economic consulting).

Most regional governments counted on regional management which had played an important role in restructuring traditional industrial regions. These in turn are closely linked to Austrian participation in EU regional politics. Regional managers are currently active in eight regions and about 25 zones. The leading organisations have developed various owner and organisational structures and about two-thirds are represented by community associations. These have various points of emphasis:

- consulting in the formulation and realisation of project ideas, particularly subsidies;
- start-ups, regional and international co-operation;
- creating key innovative projects within the limits of regional resources;
- developing and supporting co-operation and networks;

[5] Note: Austrian development laws fall into the jurisdiction of regional governments.

- devising tourist attractions and improving the quality of tourist services and facilities;
- supervision and assistance in project groups.

5.4 The Strategy of an Industrial City

In conclusion we will discuss the strategy of Austria's only true industrial city – Linz – interestingly, a city whose boom can be attributed to cultural politics. At the end of the 1980s, Linz was suffering from an antiquated state industry (VOEST, Chemie Linz) and a poor image. The city was being stifled by its reputation as an industrial centre and by the associated problems of its industrial past: poor air quality, few green areas and little to offer in the way of cultural activities. Air quality has since been improved by special measures financed by the state. The economic results of the city's businesses were significantly improved by the divestiture of corporate groups, partial privatisation, streamlining staff and by adapting the product palette to high-quality semi-finished products, manufactured using improved technologies.

The promotion of art and culture, specifically in areas which harmonise with the urban industrial heritage, such as electronic art and music, helped to transform the city's image and attract multimedia companies. Since the early 1970s an annual exhibition called "Ars Electronica" has provided artists with the opportunity to interact with new technologies. Linz also promotes culture in public spaces with open-air events such as the popular "Klangwolke" ("Cloud of Sound"). The city authorities have strongly supported cultural and political projects since the 1990s and this has greatly contributed to the new image (Goldner 2000, 33-34).

With the rise of the new medias, the Ars Electronica has become more and more important, and with some support from the city it has motivated a group of young entrepreneurs to settle in Linz (http://www.aec.at/, Knebel 2000). The city has also encouraged material investments in trade and industry with respect to site, relocation, expansion and modernisation. With the exception of new companies, support projects have been limited to particular branches (computer, telecommunications, measurement, environmental, medicinal, energy and electronic technology) (www.linz.gv.at). The success of open-air events has helped correct the exclusively industrial image of Linz, attracting visitors who can testify to the improved quality of life in the city.

References

Amt der Landesregierung Voralberg (2000) „Wirtschaftspolitik Vorarlberg". In: http://www.vlr.gv.at/, query on 12/4 2000

Ars Elektronika Center Linz Museumsgesellschaft (2000) Festival Ars Elektronika. In: http://www.aec.at/, query on 12/4 2000, Linz

Bamberger R, Maier-Bruck F (1966) Österreich Lexikon in zwei Bänden, Wien München

Bamberger R, Bamberger M, Bruckmüller E, Gutkas K (1995) Österreich Lexikon in zwei Bänden, Wien

Birner A, Delapina F, Gruber M et. al (1999) Regionale Innovationspolitik 2000. Studie im Auftrag des Bundesministeriums für Wirtschaft und Verkehr, Wien

Bundeskanzleramt (1995) OECD-LEED Länderstudie Österreich „Regionale und lokale Entwicklungspolitik in Österreich zur Restrukturierung der Industriegebiete". In: Schriften zur Regionalpolitik und Raumplanung, Wien

Faulhaber T (1992) Industrie in Österreich, Wien

Gassner S, Simonitsch W (1987) Kleines Österreich Lexikon. Wissenswertes über Land und Leute, München

Gebhardt H (1990) Industrie im Alpenraum – alpine Wirtschaftsentwicklung zwischen Außenorientierung und endogenem Potential. In: Schriftenreihe für Forschung und Praxis, Heft 99, Stuttgart

Goldner F X (1995) Die Herausforderung – Linz, Landeshauptstadt Oberösterreich. Die Ergebnisse der Sommerakademie 94, Linz

Gsteu H (1971) Länderkunde Österreichs, Innsbruck

Hammerer G, Grundler A (1995) Der Umfang der Industrie in Österreich. Studie im Auftrag der Vereinigung österreichischer Industrieller und der Wirtschaftskammer Österreich, Sektion Industrie, Industriewissenschaftliches Institut Arbeitshefte 8, Wien

Industriellenvereinigung Österreich (2000) Leitideen der Industrie – Mit Innovation und Optimismus in das Spitzenfeld europäischer Regionen. In: http://www.industriellenvereinigung.at, query on 12/4 2000

Industriellenvereinigung Österreich (2000) Steiermark, Leitideen der Industrie – Mit Innovation und Optimismus in das Spitzenfeld europäischer Regionen. In: http://www.industriellenvereinigung.at, query on 12/4 2000

Jantschgi J, Working Group STENUM (1995) Ökologische Darstellung Österreichs Industrie. Endbericht einer Studie im Auftrag des Bundesministerium für Umwelt, Graz

Knebel A (2000) GKB baut ein Viertel der Belegschaft ab. In: Der Standard vom 30.11.2000, Wien

Lichtenberger E (1997) Österreich, Darmstadt

ÖROK-Österreichische Raumordnungskonferenz (1999) Neunter Raumordnungsbericht, ÖROK-Schriftenreihe Nr.150, Wien

ÖSTAT-Österreichisches Statistisches Zentralamt (1996) Industriestatistik 1994, 2.Teil, Beiträge zur österreichischen Statistik Heft 1.211, Wien

ÖSTAT (1997) Industrie- und Gewerbestatistik 1995, 1. Teil, Beiträge zur österreichischen Statistik, Heft 1.233, Wien

Seidel H (1970) Die wirtschaftliche Struktur Österreichs. In: Wurzer R (ed.) Strukturanalyse des österreichischen Bundesgebietes. Sonderausgabe im Rahmen der Schriftenreihe der österreichischen Gesellschaft für Raumforschung und Raumplanung, Wien

Soukup K (1996) Auswirkungen des EU-Beitritts Österreichs auf den öffentlichen Unternehmenssektor In: http://europa.eu.int/comm/competition/speeches/text/sp1996_021_de.html, query on 1/25 2001

Technologie- und Marketinggesellschaft Oberösterreich (2000) Strategisches Programm OÖ 2000. In: http://www.tmg.or.at/StrategischesProgramm/3.htm, query on 12/4 2000

Weber W (1964) Die Verstaatlichung in Österreich, Berlin

Wurzer R (1970) Strukturanalyse des österreichischen Bundesgebietes. Sonderausgabe im Rahmen der Schriftenreihe der österreichischen Gesellschaft für Raumforschung und Raumplanung, Wien

URLs
http://www.inforegio.cec.eu.int, query on 04. 12. 2000, „Steiermark. Zusammenfassung des Ziel-2-Programms für den Zeitraum 2000-2006“.
http://www.inforegio.cec.eu.int, query on 04. 12. 2000, „Niederösterreich. Zusammenfassung des Ziel-2-Programms für den Zeitraum 2000-2006“.
http://www.linz.gv.at

References for the Austrian Case Study

Baumfeld G, Bratl H, Scheer G (1998) Regionalmanagement in Österreich. Eine Zwischenbilanz, Abschlussbericht im Auftrag des Bundeskanzleramtes, Abteilung IV/4, Wien
Bundeskanzleramt (1995) OECD-LEED Länderstudie Österreich. Regionale und lokale Entwicklungspolitik in Österreich zur Restrukturierung der Industriegebiete. Schriften zur Regionalpolitik und Raumplanung, Band 4, Wien
Kranawetter G (1989) Persistenz und Strukturveränderung in einem alten Industriegebiet. Fallstudie Eisenwurzen, Wien
Österreichisches Institut für Raumplanung (1996) Regionalbericht 1996, Wien
Regionalmanagement Mostviertel (1999) Tätigkeitsbericht 1999, unveröffentlicht
Schönfuss D (1991) Industriestruktur, Standortfaktoren und industrielle Veränderungen im Gebiet der niederösterreichischen Eisenwurzen, Wien
Slapansky W (1996) Musealisierung der Landschaft. Musealisierungskonzept Österreichische Eisenstraße, unveröffentlichte Studie im Autrag des Vereins Niederösterreichische Eisenstraße, Wien
Stastitik Austria (2001) ISIS data query, 2001

Newspaper Articles
Der Standard, 16.03.2000, „Eine perfekte Inszenierung der „Iron Mountains“ und chinesische Gartenanlagen
Most + Eisen, Informationsblatt des Regionalverbandes Mostviertel-Eisenwurzen, Ausgabe Herbst 2000
Treffpunkt Eisenstraße, Zeitung des Kulturparks Eisenstraße-Ötscherland, Ausgabe 1/2000

Bulgaria: Identifying Regions of Industrial Decline

Julia Spiridonova and Maria Novakova

National Centre for Regional Planning, Sofia

1 General Situation of Old Industrialised Cities and Regions

1.1 Development in the Past

The historical and political peculiarities of Bulgaria's development have invariably determined the development of industry as well as the creation and development of cities and industrial areas. Unlike a number of other European countries, industrial production in Bulgaria emerged in the early 19th century, and by 1880 the first industrial centres such as Plovdiv, Veliko Turnovo, Gabrovo, Ruse, Sliven and (later) Sofia had appeared. Enterprises were small and mainly restricted to the food industry, textiles, spirits, wine and leather articles.

Industry waned at the end of the 19th century because of competition from cheaper and better-quality West European merchandise and the exceptionally low customs taxes. This prompted the government to adopt a strong protectionist policy in an attempt to boost local industrial production. Instead of merchandise, Bulgaria imported capital from Germany, Austria, Belgium, the Czech Republic and other countries; investments were chiefly in ore mining and the processing of agricultural produce. The period 1900-1918 is characterised by a visible upsurge in industrial development.

In the 1920s industry was dominated by textiles, leather and food production. Some extraction industries were also developed (ore mining, coal mining, timber). Enterprises involved in machine-building, metal-working and chemicals were founded, located mainly in regions with qualified manpower and the necessary natural resources.

Between the two World Wars the number of industrial plants grew rapidly, totalling 3345 in 1939, with an average workforce of 30 people. Industrial plants were created in big cities, i.e. Ruse, Plovdiv, Varna, Sofia, Pleven, Stara Zagora and Gabrovo. The first more complex industrial formations appeared as well, such as the regions around Sofia and Plovdiv; the determinative factors were chiefly economic – low-priced construction plots, cheap manpower, proximity to markets etc. The problem of divergence between industrial and rural areas was just beginning.

By 1939 the processing industry accounted for 84.4 % of industrial production, with figures of 13.8 % for ore mining and 1.8 % for power generation. The industrial production per capita (expressed in US dollars) was almost ten times lower than in Germany and 20 times lower than in the UK. During the same period the city of Sofia was responsible for 23.5 % of industrial production, with other areas as follows: Plovdiv 9.6 %, Ruse 5.5 %, Varna 5.2 % and Burgas 3.7 % (Donchev 1996). In 66 years (1880-1945) the number of towns rose by 50 and urban population increased by a factor of 2.3.

The period of state economic planning (1946-1989) saw significant changes in the development and organisation of industry as well as the size and number of enterprises and the sectoral structure. The process of accelerated all-out industrialisation had major impacts on the transformation of the rural network, the formation and development of powerful industrial regions and was the origin of the ever worsening centre/periphery problem.

The country's industrial structure in the 1980s was relatively stable, without significant structural shifting, and was characterised by low fluctuations. A number of authors (Bulgarian Academy of Science 1995, Gueshev 1999) note that during the 1980s Bulgaria entered the stage of over-industrialisation, while lacking the necessary resources and markets for a sustainable and crisis-free development of the oversized industry. Meanwhile, the tertiary sector was in a rudimentary stage and was entirely absent in a number of sectors. This economic structure was highly vulnerable and did not cope with the new requirements of a market economy; the slump was severe and the country had much ground to cover to adapt to a post-industrial economic development model. Recent studies confirm that the initial over-industrialisation in the starting phase of declining output had a visible adverse effect on the depth of the 'transitional recession', but not in the later phase of subsequent recovery (Berg et al. 1999, Mickiewicz 1999).

The territorial peculiarities of industrial production boil down to the following: as a result of the quick development of industry after 1948, urban population grew rapidly. The number of towns increased from 105 to 172 between 1948 and 1970, urban population swelled from 24 % to 50 % (in 20 years the urban population grew 2.5 times). The link between industrial growth rates and urban population growth is almost functional. The 1948-1970 coefficient between the two phenomena is 0.89 (Bradistilov 1974).

Parallel with the quantitative changes in the network of human settlements there were profound qualitative shifts in the creation of major industrial centres and regions. Depending on the degree of territorial specialisation and concentration and complexity of industrial production, different forms of industrial cities and regions emerged, for example industrial nodes, centres, production complexes, etc.

The planned territorial specialisation of these formations was in compliance with the country's COMECON (Council for Mutual Economic Cooperation) specialisation, and COMECON's disintegration led to a collapse of all Bulgarian industrial regions. The centre versus periphery phenomenon also emerged, but big industrial agglomerations were, with a few exceptions, not developed significantly.

The infrastructure corridors became the main reason for a further stimulation of development along two parallel axes in northern and southern Bulgaria, leading regionalists to speak of the so-called "ring": Sofia – Pazardjik – Plovdiv – Stara Zagora – Sliven – Burgas – Varna – Devnya – Shumen – G. Oryahovitsa – V. Turnovo – Pleven – Mezdra (Vratsa-) Sofia. In the late 1970s this ring concentrated around 70 % of fixed assets, with a production share of over 75 %.

1.2 Present Situation

Basic Development Trends

A specific feature of the transition in Bulgaria to a market-oriented economy was the series of shocks and crises which the country went through, the reasons for which should be sought in a wide variety of domestic and external factors. The transition process was strongly hampered by the historical legacy, with a notably high degree of dependence (compared to other CEE countries) on the former COMECON; the disintegration of the latter has dealt a shocking blow to the Bulgarian economy. Additional hindrances were the artificial maintenance of high growth rates in the 1980s using significant external resources, saddling the country at the beginning of the transition with a huge foreign debt, and ensuring its labelling as a region considered "volatile and risky". The transition was also very much effected by delays and errors in the promotion of reforms, the lack of a clear-cut strategy and corresponding policy of industrial and spatial development, and an additional deterioration of the regional situation.

The basic macroeconomic indices characterising the condition and developmental trends of the Bulgarian economy since the beginning of transition have shown a serious slump in real GDP values, irrespective of positive fluctuations during certain periods. After a brief period of a relatively low economic growth in 1994, 1995 and the first half of 1996, the financial sector underwent a severe crisis with wide economic implications. This required the introduction of a currency board in mid-1997. The main positive outcome of the currency board was the stabilisation of the economic system, but the anticipated rapid recovery of the economy was not attained. The post-1997 GDP growth rates were lower than projected. The reason was a series of external shocks, seriously impacting on the Bulgarian economy. The Asian crisis in 1997, the Russian crisis in 1998 and the Kosovo crisis in 1999 dealt successive blows to the national economy and prospects for high-growth.

In the year 2000, after the first three successive years of moderate growth, output was still below its 1991 level; during the 1990s Bulgaria showed a negative cumulative GDP growth (-2.4 %). Although economic growth has resumed in the Bulgarian economy it continues to appear fragile, and sustained increases in productivity have not yet been realised (World Bank 2000). In 2000 Bulgaria has not yet recovered its former GDP per capita level; the current figure is about 70 % of that obtained in 1989.

In the last few years the country has achieved a balanced budget. Public debt sank to below 60 % of GDP, and after the record figures of the mid-1990s (more than 1,000 %), the CPI inflation was reduced to approximately 5 % in 2000. Meanwhile nearly 90 % of prices have been liberalised. The private sector has played an important part in the GDP growth over the past two or so years. In 1999 it accounted for about two thirds of GDP growth.

The trends in the structural proportions of the categories of essential use of the GDP are unfavourable for a sustainable upward development of the economy. The relative share of investment in GDP slumped from 18.2 % in 1991 to 11.6 % in 1998 and 16 % in 1999, and is insufficient to secure real opportunities for economic restructuring. The investment/consumption proportion within the GDP is too low to support the expansion and revitalisation of production and is much worse than in other CEE countries. In Bulgaria this correlation has constantly waned; it was 11 % in 1998 and is about 20-30 % in other countries in transition.[1]

The increasing share of the tertiary sector is a sign of modernising tendencies in the economic structure, but the low ratio between its rate and the resulting economic growth is insufficient. Since 1990 industrial development has fallen to 50 %, and because of transformations in the structure of industry there has been a rise in the share of ore mining sectors and power generation. But areas such as machine building, computer and electrical engineering and the food industry, structurally important in 1990, have now lost ground.

The lack of sustainability in most industrial regions, resulting from the rapid quantitative growth during the period of industrialisation, has been further exacerbated during the period of transition by the general social and economic crisis. The protracted crisis has sharpened regional disparities and polarised the nation. The continuation of this difficult situation is due to the country having ignored regional policy for 10 years. The main trends of changes in the economic base in regions and towns are characterised by:

- an overall slump in the level of industrialisation in all towns and regions and a narrowing of their economic base;
- the bulk of structural changes in the economic base of towns and regions has been generated by the decline of individual sectors rather than by an effective adjustment through new production;
- disintegration of past "planned" technological links and regional economic structures because of massive production close-downs, coupled with a liberalisation of imports and exports and the development of the private sector. This lessened the importance and organising role of industrial centres as concerns adjacent areas, and constricted the territorial range of industrial formations;

[1] Despite more than doubling the share of internal investments in GDP in the period 1996-2000 (reaching 16 % of GDP), it still remains below the level of the most rapidly developing economies (32 % of GDP is the average for Singapore, Malaysia, Chile and Ireland). A comparison shows that in the group of EU15 and applicant countries the level of investment in 1998 was lowest in Bulgaria (only 11.5 %). In the Czech Republic, Slovakia and Poland investment is as high as 30 % of GDP (Second Report of Economic and Social Cohesion 2001, p. 47).

- new structural relationships: services/production; the increasing share of the tertiary sector shows the modernisation of the structure of the economic base of industrial regions, but the low correlation between its growth and the effected industrial growth shows the low efficiency of the sector.
- contrast development of industrial regions and towns: continuation of the process of concentration of economic activities in big cities and an ongoing decline of industrial activities in towns.

Studying the nature of structural changes in industry and the development of industrial regions in Bulgaria we come to the conclusion that the losses in the economic base of the regions can not solely be put down to the predominant development of heavy industry. Nonetheless, these industries are largely declining industries and are the root problem of "losing" regions. In the past ten years there have been substantial changes in their development through the closure of unprofitable mines, restructuring and privatisation. Also important for the country are the processes in regions characterised by a monostructural or poorly diversified economy and a decline of the economic importance of industry (output, workforce, value added).

The Regional Development Act declared in 1999 the necessity of pursuing a regional policy and of addressing the problems in specific regions. The law introduced the so-called "regions for purposeful impact" as an object of specific support: regions for development, regions of growth, regions for transborder cooperation, underdeveloped rural regions and regions of industrial decline. The latter are shown on the map.

Organisation and Role of Spatial Planning

The radical change of the basic conditions and the inadequate normative base had a negative effect on the conditions of organisation in the territory and the promotion of growth through spatial planning. The lack of funds of regional authorities and the delayed development and adoption of new spatial planning laws delayed work on the development of new urbanisation plans. The majority of existing town and country plans were developed under the premise of no private property, absolute state rights to deal with realties and a centralised planning of investment. The industrial areas included in the plans do not satisfy investors, who prefer "greenfield" instead of "brownfield" investments. Piecemeal changes to these plans for the most problematic territorial plots led in the end to radical changes, and in some cases to an overall replacement of important and strategic urbanisation decisions. In this way, the existing urban plans are rather a hindrance in managing spatial processes.

Another very important organisational matter is the management of urban investment. Unofficial data shows that the share of private investment (not including investment in infrastructure) ranged from 70 % up to, in some cases, 90 % of all investment, i.e. a prime source of building resources. But there are no mechanisms to link private investment with the construction of infrastructure elements and the overall construction of the urban habitat. The main problems in spatial planning led to the drafting of the New Spatial Planning Act, in force from 2001.

Map: Bulgaria – Areas of industrial decline

Perspectives for Integration with the European Union

The invitation extended by the European Union for negotiations on Bulgaria's admission has had a motivating impact on development in the country, and given a push to more dynamic and goal-oriented policy efforts.

So far the integration of Bulgaria to the EU markets appears to limit its role as a supplier of goods of low value added, the manufacture of which is labour-intensive and requires the use of natural resources. The export of goods whose production requires low-skilled labour showed the highest growth rate during two phases in the dynamics of Bulgaria's trade with the European Union (1992-1995 with an export growth of 20 %, 1996-1999 with stagnation and declining growth rates in the export sector). Although the share of products requiring intensive use of natural resources in their manufacture is diminishing, the aggregate share of the resources-intensive products produced by low-skilled labour in the country's exports to the EU has stayed at about 62 % of the total since 1989 (World Bank 2000). In 2000 the share of exports to EU countries continued to grow and there was a clear shift from raw materials to consumer goods, accounting for over 40 % of total exports to the EU.

The assessment of the impact of integration processes on the development of regions experiencing serious hardships in adaption gives grounds to draw certain conclusions. The anticipated positive changes relate to:

- stimulating of regions located along the southern and south-western border and encouraging their development through programmes supporting cross-border co-operation, as well as the development of other backward regions;
- more dynamic and effective regions located along international transport corridors;
- greater access to international funds and an intensification of investment in the country as a pre-condition for successful restructuring.

However the liberalisation of trade with the EU will test the internationalisation of Bulgarian industrial plants, and will prompt local manufacturers to make changes in the quality, packaging and marketing of their products, as well as improving overall competitiveness.

2 Specific Problems of Regions Suffering Industrial Decline

2.1 Identification

For the purpose of Bulgaria's Regional Development Act, "regions of industrial decline" are municipalities or groups of municipalities characterised by a monostructural economy. These regions show a severe decline in industrial sectors, decreasing production, surging unemployment, and they are facing an economic restructuring which brings with it acute social problems.

The identification of declining industrial regions is complicated by the fact that the bulk of the country's industrial regions have some structural problems and show a slump in industry. On the other hand the boundaries which define the different types of regions are to a certain extent merely conventional. Many regions bear the characteristics of two or more types. Therefore the definition of "regions of industrial decline" is in terms of their long-term and fundamental impact on economic development and their effect on the spatial division of labour. "Regions of industrial decline" are analysed within their territorial scope, adopted by a 1999 Council of Ministers' Decree as a result of the "ordinance which sets out the criteria and territorial scope of regions of purposeful impact". Table 1 shows the indicators used for the determination of regions of industrial decline. Irrespective of their determined territorial scope for legal purposes, the problems of declining industry because of obsolete technology or lack of competitiveness are typical for most of the country's industrial regions. In this sense, the problems thus outlined are representative of regions in Bulgaria.

Table 1: Indicators for determination of regions of industrial decline (Council of Ministers)

Indicators	Value
1. Level of industrial employment	Relative share of those employed in industry from the total number of those employed in the past year – above the national average
2. Monostructure of the economy, dominated by industrial sectors in decline	Relative share of those employed in the sectors coal mining, ore extraction and metallurgy, metal casting, chemicals, textiles, from the total number of those employed in industry in the past year – above 50 percent
3. Level and dynamics of unemployment	Mean annual unemployment rate for each of the past 3 successive years – above the national average for the respective year
4. Dynamics of industrial unemployment	Negative growth of those employed in industry for the last 6-year period – above the national average
5. Industrial decline	Drop of industrial output above the national average

Source: 1999 Council of Minister Decree, own compilation

There are nine regions labelled "regions of industrial decline", as well as other regions where the principal field of employment is in coal mining, the energy sector or ferrous metallurgy. At this stage such regions have not been classified as regions of industrial decline because the government is convinced that the coal and ferrous industry will remain competitive, and that the risk is low that a critical situation will emerge.

2.2 Situation and Problems in Regions with Industrial Decline

Within the defined spatial area, the regions of industrial decline cover 9 percent of the country and have a combined population of 530,000 inhabitants, or 8.3 percent of Bulgaria's total population. The industrial sector is the main source of employment. In some regions the level of employment in industry is above 60 percent.

The monostructure of the economy is a typical characteristic. The share of old industrial sectors in manufacturing varies from 37 to 80 percent: 44.3 percent of coal mining, 65 percent of non-ferrous metallurgy, 28 percent of ferrous metallurgy, 12 percent of the textile industry and 7 percent of the chemical industry (mainly basic chemistry) are concentrated in these regions.

Specific problems of development in regions with industrial decline are: the drop in industrial output, ranging between 40 and 70 percent from 1990-1998, constituting one of the highest declines in the country. The economic base of the cities has been narrowed, and certain production lines have been shut down or greatly reduced. The shift to alternative activities is weak and the emergence of new small and medium-size enterprises is severely restricted. The general drop in economic activities has had a negative impact in many different areas, and problems are additionally aggravated by the absence of the necessary conditions for rapid rehabilitation of industrial sites. The sole "positive" effect of shrinking production is to be found in reduced pollution emissions from manufacturing facilities.

A further problem is the high level of unemployment. During the 3-year period under review (1996-1998) the level of unemployment exceeded the national average by 50 %. In the overwhelming majority of regions and cities it exceeds 15 percent and varies in the range 20 to 46 percent, whereas the average for the country as a whole is about 14-16 %.

The local impact of any reductions in employment is viewed in terms of both the direct and indirect effects. The latter are largely a result of a decrease in local purchases of goods and services, both by the mining companies and by miners and their families as a consequence of the severe curtailing of their disposable income.

The concentration in some areas of large energy producers or metallurgy enterprises has determined some regions as ecologically depressed. In recent years there has been a marked reduction in harmful emissions due to the fall in production. However a number of other forms of pollution – that of soils, water and air – still represent a grave environmental problem for these regions. For example, even though in Pernik the level of nitrogen oxides has diminished from 6,200 to 2,213 tons, that of sulphuric oxides from 15,773 tons to 12,538 tons and that of carbon oxide from 18,951 to 5,137 tons (1990-1998), the region is still one of the country's pollution hot spots. The ecological problems in some areas (Madan, Rudozem, etc.) have led to serious health risks and high disease coefficients.

The cities, representing the core of the regions of industrial decline, are predominantly small and medium-sized. They are sensitive to the effects of recession, particularly to the industrial crisis, and no major changes are expected without outside help. Because of the worsening economic situation and the poor technical, social and market infrastructure, the situation of small towns over the last decade has been precarious and their rural influence perimeter has been narrowed.

3 Explaining the Problems of Regions in Industrial Decline

The industrial or economic decline of the regions has been linked to the macroeconomic situation in the country as a whole, the international environment and the ongoing process of globalisation. Within the framework thus described, one can find differences in the development of various regions; this is the result of specific regional factors.

Table 2 shows the problems of the regions of industrial decline. The characteristics of these regions indicate some of the reasons for their decline. A more complete presentation of reasons would help in the formulation of necessary action plans for new regional policy. Therefore some major regional and local reasons for the difficult situation will be specified.

Local Economic Structure

Structural factors play an important role in the troubles faced by industrialised regions and cities. Although almost 10 years have elapsed since the onset of the transition period, the economic structure inherited from the years of central state planning is still an active factor in the depression or decay of regions and cities. The combination of monostructure with a high relative share of declining industries or the predominance of depressive sectors contributed to the genesis of these changes.

In the monostructured regions, where large industrial enterprises have strongly reduced their activities, one can see that the economic potential of the region has been gravely undermined, and much time and effort will be required to restore and develop positive structures. The economic crisis has resulted in a process of "passive" structural change with a natural "demise" or "down-sizing" of enterprises, and this has rarely been accompanied by "active" factors such as decisions on the closure of uncompetitive enterprises. The lack of specific measures in the field of structural policy to counteract the negative tendencies has had an impact on the regional horizon.

Although delayed, the structural reform process is coming to an end. The accelerated stages of liquidation and insolvency proceedings are leading to significant reduction of losses in the productive sector. By 1st July 2000, 189 public companies showing a loss had undergone, or were in a process of, liquidation.

Investment Activities

The very limited investment resources in the country have greatly delayed the processes of restructuring and economic recovery, as well as postponing the establishment of alternative activities. One of the reasons for this is that public finance is controlled by a Currency Board that strongly limits domestic public investment expenditure. The budgets of local authorities and their investment opportunities are very limited, and this further decreases the physical stock of capital in certain regions of the country.

Table 2: Specific problems of regions in industrial decline

Problems	Characteristics
Economic problems	– decreasing production; – low diversification of economic activities; – low degree of economic competitiveness; – de-industrialisation;
Problems in mentality	– low degree of entrepreneurship; – reliance on the state for decision-making – "outdated thinking" in the population; – low ability for re-qualification;
Problems in human resources	– demographic decline; – negative balance of net migration; – youth migration for economic reasons and lack of motivation – low level of educated labour force;
Infrastructure problems	– lack of infrastructure for economic support; – obsolete basic and social infrastructure;
Social problems	– high level of unemployment; – shrinking average income; – decreasing local purchases of goods and services;
Ecological problems	– highly degraded soils, air and water pollution; – devastated landscape, disturbed urban patterns, health risk; – "positive" effect of less production – reduced pollution emissions;
Investment problems	– low attractiveness for foreign investors; – lack of investments for restructuring;
Human settlements problems	– narrowed economic base in cities and municipalities due to economic decline; – urban and rural degradation;
Spatial planning	– lack of new urban plans; – need of rehabilitation of terrains and construction of modern industrial sites.

Source: own compilation

Considering the difficulties associated with the unfavourable external environment (e.g. the period of armed conflict in neighbouring countries and the effect on investment of the Russian crisis), Bulgaria has achieved great success in consistently attracting new flows of capital. Over 52 % of this investment has taken the form of joint ventures, greenfield investment, reinvested earnings and credits, while 40 % is attributed to privatisation and 8 % to capital market investment. Three European Union member countries, Germany, Belgium and the Netherlands, are, with an almost two-thirds share, the biggest investors in Bulgaria. Expectations and confidence in the business sector have gradually and consistently improved since early 1999. By 2000 the Business Confidence Indicators for Bulgaria were growing steadily. However, stagnant domestic demand continues to dampen business confidence (Bulgarian Common Country Assessment 2000). In comparison to the countries of central and south-eastern Europe, Bulgaria's investment inflows are relatively low. Romania, Croatia, the Russian Federation,

Hungary and the Czech Republic have managed to attract more foreign direct investment (World Investment Report 2000).

Foreign capital has highly selective spatial behaviour in the country, and confirms conclusions about the influence of economic, geographical, functional and demographic characteristics of regions at the national and international-European level and the polarising pattern of regional growth (Petrakos 1999). The influx of foreign investment is mainly oriented towards more attractive regions showing comparative advantages. Only 4 % of the foreign direct investments have been spatially addressed to the studied regions, which account for nearly 10 % of the country's land area and population.

Local Infrastructure

The level of infrastructure development is one of the underlying reasons for the decline of a given region. The regions have a technical infrastructure which is not appropriate to meet today's requirements and impedes the ability of the regions to react rapidly to new innovative manufacturing opportunities or to efficiently expand existing facilities. Equally, those municipalities with a better infrastructure base demonstrate greater flexibility; their development efforts are not oriented towards keeping an obsolete structure intact, but focus on new opportunities and horizons.

When analysing the level of infrastructure in the individual regions, one finds great differences in levels of completion. This is above all connected with the concentration of material facilities in the past, and measures taken for their maintenance and rehabilitation in the period of transition. As mining enterprises were previously able to provide support to the municipalities, these regions can at least claim an infrastructure that meets the daily services and demands of the population. However, from the point of view of infrastructure acting as a catalyst for processes of innovation and rapid adaptation or restructuring, the available infrastructure is clearly outdated and badly maintained.

Human Resources, Local and Regional Capacity

A typical feature of these regions is the limited profile of the skills level and educational background of the labour force, this being especially true of regions dominated by small towns. The long years of one-sided occupation in these regions and the lack of retraining measures has intensified the negative inertia factor in their development, and has also had an impact on the level of entrepreneurship, flexibility and adaptability to new conditions.

Achievements in regional and local development depend significantly on actors' ability to recognise their potential, to articulate demands and mobilise for collective action. In Bulgaria past experience has confirmed that there is a likely correlation between the level of municipal development and perceptions of capacity (Human Development Report 2000). For now, it seems clear that local capacity needs to be developed and more encouragement given to overcome existing problems in the regions. Such problems are:

- the lack of strong private or public institutions which can drive forward the process of transformation into more sound economic structures. There is how-

ever some commitment shown by key people in some organisations; this might be strengthened by adequate support measures from institutional structures;
- apart from objective difficulties linked to the receipt of bank credits, only a very small portion of existing and potential local entrepreneurs have the necessary information and training to develop successful business plans and marketing strategies. The lack of such services is one obstacle in the development and expansion of entrepreneurship;
- the low degree of involvement of some organisations in the process of local and regional development seems partly a reflection of the gulf between public position-holders on the one hand and private business circles on the other. Both sides are still not aware of the interlinkage and positive impact of joining forces in a civic society; there is a low representation of the private business sector or SMEs in the emerging local development assistance bodies;
- insufficiently developed business and professional training and a low degree of know-how in project preparation and management.

Studies of the level of entrepreneurship in regions of industrial decline and in backward regions, as measured by the density of companies, show that it is two times under the national average. Following the conclusions of the 1996-1999 Report on SMEs, poor entrepreneurship should not be exclusively attributed to a low entrepreneurial culture or specific psychological characteristics of the locals, but rather arises from the economic environment and the infrastructure, i.e. physical and institutional factors. The authors also note a delay in the process of creating new SMEs in the region and a much lower occupation share (19 %) as compared to the 1998 average (44 %).

EU Integration

Various surveys form different conclusions as to the impact of the current and future models of trade on the development and productive structure in the regions and countries of CEE. The Second Cohesion Report confirmed what many research projects have suggested, that enlargement will produce longer-term profits for both the EU 15 and the candidate countries, with variations according to the particular sectoral structure and trade relations. On the one hand there will likely be profits for the 15 EU member-states and regions that are specialized in investment goods, high-productivity industries, businesses and financial services. On the other hand there will be competitive pressures on EU producers of labour-intensive goods (e.g. textiles, footwear), certain agricultural products (e.g. grain, vegetables, fruit, cattle, pig farming) and products with a low level of technological sophistication (e.g. printing, chemicals). For the CEE countries and regions, some of the research predicts competitive advantages arising from low labour costs (in areas such as textiles, steel and chemicals), but also disadvantages (especially for SMEs) given the combination of greater competition and the additional cost of compliance with EU regulations (Second Report on Economic and Social Cohesion 2001). Results from other studies give rise to uncertainty regarding the possibilities of convergence, determining that East-West trade relations have a predominately inter-industry character (Landesmann 1995, Petrakos 1999, Mack and Jakobson 1996). West European countries tend to specialise in knowledge-based products requiring intensive research and development (R&D), while central and eastern Europe tends to focus on labour-intensive and resource-intensive products. As a result, trickle-down effects are assumed to issue from core regions to the periphery.

In regard to Bulgaria, another conclusion can be confirmed. The liberalisation of trade with the EU will test the internationalisation of Bulgarian industries. As a result, more sensitive monostructured regions and regions with structural weaknesses and low adaptation are expected to suffer directly from the openness of the economy. The activity of most SMEs in studied regions is strongly influenced by fluctuations in the local economy, and their potential for expansion and income generation is entirely dependent on other companies in the region. With the economic restructuring of regions of industrial decline and a possible loss of jobs in major companies, there is little expectation that the policy of encouraging the creation of new SMEs will lastingly resolve the employment problem.

In a monostructural economy most of the workforce is employed in a small number of companies with a narrow business profile, and accumulated experiences can hardly be used in the future. The existing policy of encouraging the formation of small privately-owned businesses by granting purpose-oriented funds as initial capital is only a short-term policy. In the long-term it will prove necessary to develop and implement strategies and programmes which fundamentally change the environment and capacities of local human resources (Small and Medium Size Enterprises 2000, 83, 86, 91, 92).

4 Strategies for Further Development

4.1 Definition of Development Aims

The set of aims forming the strategic development of regions of industrial decline, irrespective of certain regional peculiarities, can be determined for Bulgaria's conditions as follows:

- creation of a vital economic infrastructure, operating efficiently under market conditions with possibilities for a dynamic reproduction;
- protection and nurturing of regional human resources;
- employment generation;
- creation of stable and competitive infrastructures;
- ecological urban restructuring following the decline or closure of major industries;
- improvement of well-being and general quality of life;
- measures to protect and improve the environment.

Fulfilling these aims will require different measures and interference in regional development depending on the characteristics of the region and the resources which can be mobilised for implementation. The improvement of regional and local governance or, more exactly, the attainment of good governance, is not included in the above aims, but is of course of major importance for their implementation. In this respect it can be raised to the rank of a separate objective.

4.2 The National Plan for Regional Development

After a long period of state policy which prioritised the attainment of macroeconomic goals, there has been a general turn over the past few years to the problems of regional development and the means of implementing regional policy. The motivations for this are to prevent the further process of degradation of regional structures and also to focus on regional factors in generating economic growth.

In 1999 the Regional Development Act was adopted. The restrictions imposed by the International Monetary Fund (IMF) on taxation and on the creation of budget or extra budget assistance funds hindered the attainment of the initial goals of the Act, notably special emphasis on the development of certain regions and the formation of the Regional Development Fund.

The Regional Development Act, as the main instrument of regional policy, determines the National Plan of Regional Development for a period of seven years. Regarding the distribution of state subsidies, the 2000 Budget Act says that municipalities should give priority to projects included in regional development plans. This restriction was included as currently only a few municipalities make capital expenses on the basis of a regional strategy and development programmes. The projects in the National Plan are secured not only by budget financing, but also by various other fiscal sources, including existing extra budgetary funds, EU accession funds and others.

The plan is set up using a combination of top-down and bottom-up approaches. This means, briefly, that regional and district development plans are based on municipal strategies and serve to form the National Plan. Each planning level calculates its own strategic goals and assesses the regional and national effect as well as determining the relative importance of priorities and projects. A consensus over project priorities is reached by adopting municipal strategies from municipal councils, district plans from the district councils for regional development, and the National Plan from the National Regional Development Council. There will also be an annual updating of the plan in line with the described technology.

The elaboration of the first National Regional Development Plan 2000-2006 showed a series of deficiencies. The administrative capacity for developing local strategies is insufficient, and a necessary reorientation of administrative cadres in municipalities to deal with issues of business, entrepreneurship and marketing is difficult und will require much time. There is a lack of skills in strategic planning, dialogue, partnership development, etc. There are not enough mechanisms to secure the diversity of support actions required for important measures and to realise projects in municipal strategies and programmes.

The policy measures of regions in industrial decline represent a separate section of the plan. The measures envisaged in the National Plan for all regions are systematised in terms of priorities, as follows:

- economic development, including programmes on industry, agriculture, tourism, labour, business infrastructure, SMEs, technological development, etc.;
- technical infrastructure, comprising programmes and measures for development of the individual systems;
- social infrastructure, including programmes on education, health care, social care, culture and heritage, sports and youth activities;
- public works;
- protection of the environment, comprising a range of programmes, including the Programme on Environmentally Safe Rundown of Mining and Extraction Activities;
- cross-border co-operation;
- human resources, comprising the programmes on education and training;
- research and design, an important part of which is the programme on research and development activities.

The National Plan provides a set of recommendations, and therefore there is no mandatory requirement that all projects incorporated therein have guaranteed funding. The relative share of guaranteed funding for implementation of projects in 2001 is highest for the regions of growth, with 64 % guaranteed funding, followed by the regions for cross-border co-operation (58.9 %) and the regions in industrial decline (58.6 %). The total amount of guaranteed financial resources for investments in the regions of purposeful impact is 52.5 % of required funds. The major sources of financing are the state budget, local budgets, the EU programmes PHARE, ISPA, SAPARD, other donor programmes and credits backed by state guarantee.

4.3 Strategies to Tackle Structural Deficits

The implementation of policies to overcome the processes of crisis and to attain positive social and economic transformations as well as accelerated economic growth is a major task in development, both nationwide and regionally. The realisation of policies is ever more interpreted as a process of active partnership at all levels of administration.

The Role of Municipalities

Municipalities are the basic level of self-government in the country, while districts (oblasti) are a level of state administration. The principal instruments by which municipalities may influence local development are: municipal development strategies, municipal and urban spatial plans, procedures for launching new business activities, the issuing of building permits and allocation of plots and zones for business activities, the application of certain preferential conditions assigned to them by the law (the Public Procurement Law, the Law on Small and

Medium-size Enterprises, the Law on Public Property, etc.) and, lastly, the use of local budgets and funds for privatisation and environmental protection (set up by the municipalities and spent under specific decisions of the Municipal Council). Experience has shown that these opportunities are not always used to favour local development, with factors such as bureaucratic sluggishness, lack of knowledge on the legal opportunities etc., proving a hindrance.

Furthermore, Bulgarian municipalities face a serious gap between their obligations and available resources. The revenue side of municipal budgets does not correspond to expenditures, especially in terms of items and budget lines. The shortage of financial resources and the limited opportunities for increasing their own incomes and independent governance are some of the most serious obstacles stopping municipalities from contemplating more feasible financial projects.

In the past two years (1999-2000) the principle of strategic planning has firmly established itself in the municipalities. Municipal development strategies have a dual purpose. On the one hand they form the basis of the pyramid of regional planning and development, thus realising a top-down approach in regional planning. On the other hand the municipalities are sufficiently independent to set political priorities and pursue their own policies of regional development. The municipality is free to seek and realise measures, acts, and projects, outside those in the National Regional Development Plan, while realising the same aims of development.

Strategies

The policy for the *big industrial centres and cities/towns* focuses mainly on the restructuring and internationalisation of the regional productive system. This is realised by fostering SME development, attracting foreign investment, enhancing technological potentials, supporting environmental improvements for enhanced quality of life, rehabilitating zones for other economic activities, and by the rehabilitation and reclamation of polluted and contaminated areas.

In the *monostructured industrial areas,* comprising smaller municipalities and areas, the objectives of the strategies are usually related to:

- enhanced diversification of the local economy;
- support for local entrepreneurship and attracting external investments;
- support for human resource development, mainly through training programmes and thus increased employability;
- regeneration and improvement of small-scale infrastructure in towns;
- creation of more attractive town environments for enterprise investment and better social mix.

Tourism is perceived as a development alternative in all regions and districts, but especially for areas suffering economic decline and a high level of unemployment. Indeed studies have shown that all areas in Bulgaria have a potential for tourism development. Municipalities with a comparatively high level of tourism development are not in the list of areas most affected by economic crisis. The development of *basic infrastructure* in terms of economic development support and the creation of favourable conditions for living is strongly present in the strategies.

The role of infrastructure is clearly defined as a condition of, and motor for, business development and attractiveness of the region.

There is a determined and heightened interest in Bulgarian regions as regards *cross-border co-operation.* The establishment of lasting and mutually-beneficial contacts with neighbouring border municipalities and regions is frequently proposed, as is the organisation of joint economic and social activities. For the regions Simitli and Pernik, the development of links with the Republic of Macedonia and Republic of Greece have enabled an enhancement and real activation of tourism (Simitli) as well as trade exchange and new investments (Pernik).

Despite some effort, the correct mechanisms to secure realisation of projects (included in local strategies and programmes) have not yet been found. In any case, the development of municipal strategies is a pre-condition for official acceptance of specific projects when seeking funding. Such strategies for municipal socio-economic development must also be formulated when grants are sought from various funds, schemes and other donor programmes.

The limited possibilities for foreign investment in these regions should always be borne in mind, while the promotion of *endogenous potential* should be at the basis of development plans, with emphasis placed on the transfer of technologies, business councils, education and the optimal use of local resources.

In the same way as the state is involved as a partner in the implementation of municipal strategies, the municipalities are present as partners in the implementation of national strategies that may influence the development of industrial areas. It is worth noting once again that the regions designated as suffering industrial decline do not cover all areas and cities in the country facing serious problems in their industrial development and needing restructuring and specific development policies.

National Strategies and Programmes

Here it is necessary to mention the national strategies and programmes formulated for a variety of fields. The *development of industries* is supported by actions such as the development of technology parks, improved industrial infrastructure, business development aid for selected sectors, quality certifications, organisation and management improvement schemes, financial and credit instruments for export promotion, etc. The *development of tourism* is fostered by improvements in tourist infrastructure (accommodation facilities, eco-tourism trails, cultural sites, tourist information and visitor centres etc.), training for those working in tourism, support of marketing and advertising, introduction of information technologies, alternative tourism projects, events etc. There exists a wide range of instruments for promoting the *development of SMEs*, such as financial and credit assistance, business incubators, consultancy services, information and technology transfer centres, quality certifications, introducing environment-friendly technologies, simplifying the bureaucracy imposed on small businesses (especially in the areas of registration and taxation), the setting up of a micro-credit scheme to make access to credit easier and cheaper for small companies, simplification of the administrative procedures faced by SMEs, etc. Six business incubators are to be created in the regions of industrial decline as part of the PHARE Programme. Currently a project to enhance the capacity of established Regional Development Agencies as well as to create some new ones is underway, also sponsored by PHARE.

The active labour market policy focuses on unemployed persons, employers, employees, and social partners, and aims to establish favourable conditions for increased employment and reduced labour market pressures. The share of funds going towards active labour market measures has been increasing, and the following measures and programmes have been implemented in the country: programmes for reducing unemployment and special programmes for unemployed youth, encouraging employers to hire long-term unemployed and young people; assistance in relocation to find work in a different region; the creation of employment associations to provide work for unemployed persons; creating new jobs through the development of social and technical infrastructure projects; encouraging employers to hire unemployed people on a part-time basis, and increasing employment by encouraging employers for the first five employed persons. In addition to these measures there are also programmes which facilitate the transition from school to the job market, such as assisting youngsters from orphanages to find work after leaving secondary school. This programme will continue to be implemented as a result of an agreement between the Ministry of Labour and Social Policy and the Ministry of Education and Science. Further measures include vocational orientation programmes, projects to encourage the employment of vulnerable groups (such as the disabled), programmes for literacy or to provide qualifications, and a national framework programme for the social and economic integration of the Roma people.

Development of entrepreneurship programmes offer assistance and encouragement by the establishment of business centres throughout Bulgaria; self-employment is stimulated through the "Start your own business" programme; training and financial support is given to start businesses in the unemployed persons programme, etc.

Sectoral, regional and local employment programmes are elaborated and implemented to develop the possibilities of *job creation.* Six industrial development programmes have been developed and implemented for areas with growing unemployment as a consequence of the restructuring of the local economy; these are Chiprovtsi, Lom, Karlovo, Pernik, Smolyan and Stara Zagora. The implementation of projects envisaged in the economic programmes will ensure the creation of 4,757 new jobs. Funding for these programmes is provided by allocations from the state budget and from EU pre-accession funds (under the programmes SAPARD, ISPA and PHARE) as well as from internal and foreign investors.

With a view to funding employment promotion activities a Social Investment Fund was set up in 2001. It will enhance opportunities for the financing of projects in the field of employment outside the scope of the currently operating Fund on Professional Qualification and Unemployment.

Pernik and Simitli – Contrasting Basic Conditions for Structural Change

Irrespective of the similarity of identification indices, the problems of the regions of industrial decline show differences from region to region. The two localities of Pernik (west of Sofia) and Simitli (in the south-west of the country) have many of the social and economic problems that coal-mining areas face in other parts of Bulgaria. One has a larger and more diversified employment structure, while the other is a small, rural locality with heavy dependence on the mining industry. They are ideal for illustrating certain problems and to help design approaches that can be applied to other parts of the country.

The diversified economic base of the industrial complex of Pernik may have assisted the area in maintaining an unemployment rate (16 %) that is below the national average (18 %). The city was deliberately developed as an "industrial complex" and became a major centre for steel production, machine construction, textiles and mining. By the late 1980s almost half of the workforce was engaged in these activities. Economic difficulties since 1990 have resulted in a loss of output and a steady rise in unemployment. While the strong industrial base of this region means there is a significant infrastructure of physical and human assets, the territory still needs to adapt to grasp future opportunities, and continued effort must be made to attract new investment. However there are already indications that the area is indeed capable of capturing the interest of international investors, for example with joint venture activity in the clothing and textile industries.

The area of Simitli provides a contrast to Pernik in scale, industrial structure and location. The locality is mainly rural and industrial development is limited, with some activity in ceramics manufacture, in small-scale wood processing and in the production of building materials. A major employer is the mining industry. Any significant reduction in the mining workforce at Pirin would have an especially severe social impact on Simitli. The current economic structures show few indications of strength or dynamism, and prospects for job creation in the absence of new special measures are likely to be limited. The area is presumably not attractive to substantial new industrial investment from outside, implying that it must instead seek the maximum possible benefit from internal developments in the indigenous SME sector, as well as adopting strategies to exploit the area's potential for tourism, both in winter and summer. The unspoiled mountain landscape, along with a basic level of services for winter tourism, offers a starting-point for a focused sector development programme.

Julia Spiridonova and Maria Novakova, National Centre for Regional Planning and Housing Policy, Sofia

The creation of local investment and guarantee funds to finance employment programmes is also envisaged. In order to assist the structural alignment of the economy and to diminish social tension caused by restructuring and privatisation of the steelworks and the mining industry in the three major regions affected by these changes (Sofia/Pernik, the Rhodope area and Bourgas), the implementation of a Programme for Employment in the Steelworks and the Mining Industry was begun in the period 2001-2003. Part of the programme is financed by funds allocated under PHARE and other donor programmes. An important element of the programme is the introduction of a system for grant assistance that will support:

- local initiatives that promote opportunities for permanent employment of the labour force in those regions affected by the closure and restructuring of steelworks and mining enterprises;
- initiatives to promote employment through projects for development of the municipalities, the social welfare system, infrastructure and the environment, thus helping to raise living standards and improve the social environment, while creating conditions for permanent employment and preventing long-term unemployment;
- initiatives for training and retraining to encourage the unemployed to return to the labour market in the areas affected by the closure and restructuring of steelworks and mining enterprises.

The programme aims to provide opportunities for employment and training to those discharged from the steelworks, the mines and related industries, as well as members of their families who live in the identified municipalities. These opportunities are provided by projects developed and managed by local organisations and financed by donations from the European Union. The projects will create employment for 1,000 people. They will also provide opportunities for the training of 4,200 people, as well as giving specific assistance to more than 1,000 entrepreneurs. Types of projects that may be selected are projects that support initiatives for employment or the rehabilitation of the environment

4.4 Strategies to Improve Organisational Deficits

Up to the present time the performance of institutional and management functions was marked by deficit, distortions and inadequacy, with, for example, sharp contradictions due to the parallel existence of legal and "grey" economies, inefficient interaction between institutions of public, business and civil society, high dependence on central budget financing, a lack of co-ordinated policies and poor skills for the attraction of investors.

The smooth functioning of the business sector requires a certain degree of speed and flexibility. The public authorities have difficulties in always this, and they seldom have the necessary resources to assist local businesses. The realisation of measures aimed at founding special regional institutions (Regional/Local Development Agencies and the like) is part of a series of strategies to improve matters.

The principal functions of the Regional and Local Development Agencies are as follows:

- providing information, realised through both passive/active dissemination;
- marketing of the location, advertising and presentation to external investors and customers;
- consultation and advisory functions;
- lobbying on a regional scale.

The establishment of Regional and Local Development Agencies creates certain difficulties in the smaller municipalities and cities, with regard to employing skilled and motivated staff who have the capacity to identify with customer groups. An element of critical importance for the business incubators is the provision of efficient business services, i.e. a well developed package of advisory services. This impinges again on requirements for appropriate staffing, and the training of future entrepreneurs. One of the solutions to this is the provision of financial support for business centres, business incubators and the Regional/Local Development Agencies by the municipalities or through a national programme, according to the financial resources of the small municipalities. It is important that the support should be directly linked to the quality and quantity of the services that these structures provide to the small and medium-sized enterprises and towards the development of entrepreneurship.

In regard to institutional practices, measures are also being undertaken to strengthen *co-ordination and co-operation* between different political-administrative levels (municipal, regional, national), between different fields of policy on the same administrative level, and between different cities, towns and regions. The process of investigating and implementing development strategies and plans, as well as of the projects for local or urban development, contributes greatly to the attainment of objectives.

Local Development Councils have been established in all municipalities. These councils comprise representatives of the respective municipality, the district, decentralised governmental services, private business and the non-governmental sector. The task of the councils is to elaborate a common development vision and to support collaboration and partnership between the various actors in resolving common problems and protecting mutual interests.

In conclusion, it is worth noting that it is still difficult to pin-point exactly all those trends which will contribute positively to the development of industrial regions in decline. The limited resources and the high inertia of regional development means that the implementation of plans and strategies for regional development is generally a slow process. Unrealistically high expectations can lead to a renunciation of the applied policy before it has had a chance to bring about the desired results.

References

Agency for Small and Medium Size Enterprises (2000) Report on small and medium size enterprises 1996-1999, Sofia
Berg A, Borenstsztein E, Sahay R, Zettelmeyer J (1999) The evolution of output of transition economies: Explaining the differences. IMF Working paper No 99/73
Bradistilov D (1974) Territorial planning, Sofia
Bulgarian Academy of Sciences (1995) Bulgaria at the beginning of XXI century, Sofia
Bulgarian Common Country Assessment (2000), UNDP, Sofia
Donchev D, Karakashev H (1996) Physical and socio-economic geography of Bulgaria, 4th edn, Sofia
Drinov D (1997) Geography of Bulgaria. Physical Geography and Socio-Economic Geography, Sofia
European Commission (2001) Second report on economic and social cohesion, Luxembourg, pp 73-76
Final report of the Phare project (1998) Bulgaria: Pilot closure of an uneconomic coal mine, Ministry of Economics, Sofia
Gueshev G (1999) Problems of regional development and regional policy of Bulgaria. Bulgarian Academy of Sciences, Sofia
Human Development Report (2000) Bulgaria 2000. The municipal mosaic, UNDP, Sofia
Mickiewicz T (1999) Convergence versus rapid deindustrialisation. Restructuring of employment in Central Europe. Paper presented at the Workshop on Regional Development and Policy, Bonn
Landesmann M (1995) The patterns of East-Europe integration: catching up or falling behind. In: Dobrinsky R, Landesmann M (eds.) Transforming economies and European integration, Aldershot: Edward Elgar, pp 116-40
Mack R, Jakobson D (1996) The impact of peripherally upon trade patterns in European Union. In: European Urban and Regional Studies, vol 3: 364-9
Ministry of Labour and Social Policy (2001) National Plan for Employment Actions in 2001. Final Draft, Sofia
National Centre for Regional Development and Housing (1998) Setting out the criteria and territorial scope of regions of purposeful impact, Sofia
National Human Development Report (1999) Bulgaria 1999. P.1: Trends and opportunities for regional human development, Sofia, UNDP, 1999. In: http://www.undp.bg
National Plan for Economic Development of Bulgaria 2000-2006. In: http://www.government.bg.
National Plan for Regional Development of Rep. Bulgaria 2000-2006. In: http://www. government.mrrb.bg.
National Statistical Institute (2000) Bulgaria'98. Socio-economic development, Sofia
Petrakos G (1999) Patterns of regional inequality and convergence-divergence trends in transition economies. International Conference on regional potentials in an integrating Europe. Regional Studies Association, Bilbao
Petrakos G (1999) The spatial impact of East-West integration in Europe. In: Petrakos G, Maier G, Gorzelak G (eds) Integration and transition in Europe: The economic geography of interaction, London Routledge, pp 58-79
Vaknin S (2001) Bulgaria's economy and the challenge of accession, CER vol 3, No 18. In: http://www.ce-review.org/01/18/vaknin18.html
World Bank (2000) Bulgaria. Country Economic Memorandum. The Dual Challenge of Transition and Accession. In: http://www.government.oficial_docs/index/htmlg
UNCTAD (2000) World Investment Report, Geneva

Czech Republic: Towards the Reclamation of Derelict Land

Jaromír Stejskal[1]

Institute for Spatial Development (ÚÚR), Brno

1 Introduction

One precondition for the successful rebuilding and modernisation of economies is the identification and correct handling of processes of industrial change and decline. This includes the return and reclamation of used land that has been damaged over the years. The post-communist economies of Central Europe are developing sophisticated strategies to ensure modern free-market economies, capable of competing on the international stage, and which will provide a high quality of life for their citizens. If damaged land is not converted into a productive asset, but remains a potential health hazard and visual blight, then positive economic change is thwarted.

Dereliction of industrial land and buildings is an inevitable consequence of normal business and product lifecycles. It is particularly concentrated in regions suffering structural economic decline, and occurs as manufacturing industries and companies die, change their locations, develop new products or exhaust natural resources. In the future, better regulation of industry and business should go some way to mitigating these processes. The common motivation for remediation is the elimination of contamination and other health hazards, making land safe for reuse. Damaged land may also be a visual scar and thus come to depress local people, as well as lowering land values and discouraging investment. Foreign investors are especially deterred when potential investment opportunities are associated with health hazards. In short, derelict and damaged land has a negative impact on the image of such regions: it provides visual evidence of an out-dated economy, lacking in dynamism. Moreover in some locations reclamation of derelict land may be the only way of creating flat developable land suitable for industry. Land reclamation should therefore not be confused with the environmental problem of treating contaminated or dangerous land. It constitutes both an economic problem and an economic opportunity.

---

[1] The discussion of the problem of derelict land and recommendations for action are mostly based (with the occasional direct quotation) on a study by Arnold and Griffith (2000).

The original owner of a derelict site may often not be in a position to redevelop the land (whether through insolvency or general lack of resources), so reclamation is unlikely to be carried out by the private sector or by unprofitable state enterprises. The Czech Republic has to face the fact that a national land reclamation effort needs to be funded primarily with public money. This, of course, is a political decision and depends for its realisation on shifting national priorities.

The State's interest and motivation for commissioning the preparation of this strategy is the need for serviced land (extensive, high quality sites), mostly for foreign investors, as well as the removal of visible dereliction (on-site, adjacent to sites or in "ghost towns") that symbolises the decline of local economies and their unsuitability for modern international businesses. The priority is to help local, regional and central governments meet their objectives, by increasing the supply of attractive land in older industrial regions of the country that are suitable for foreign manufacturing investment. General guidance has to be given with regard to the wider national issue of removing hazards and improving living conditions and opportunities.

Land reclamation should not be confused with the removal of contamination. Recovered contaminated land may not be usable and may still contain derelict buildings. The issue here is not simply one of improving environmental conditions. Of course, health hazards caused by derelict buildings and neglected land can and must be dealt with. But the prime consideration in the reclamation of land is to create a usable asset, and to remove visual blight or pollution as a means to encouraging economic activity and investment. It must be recognised that land reclamation needs public funding. Dereliction causes market failure, and the normal rules of supply and demand are insufficient if the cost of damaged land is greater than the normal market price of usable land. The following section outlines the situation of damaged land in the Czech Republic, and section 3 points out affected regions undergoing structural change. The article concludes with recommendations for future policy (section 4).

2 Damaged Land in the Czech Republic

2.1 Background

Older industrialised economies usually have tracts of damaged land which are unusable without serious or expensive reclamation projects. There are various causes for this, and at least five can be identified in the Czech Republic:

1. obsolete and exhausted basic industries;
2. active and abandoned extractive industries;
3. redundant military bases; and
4. obsolete manufacturing industry.

Point one is explained by long-term historical considerations. In common with many older industrialised economies developed in the 18th and 19th century, the

Czech Republic suffers from damaged land arising from coal extraction and metallurgy (iron and steel manufacture). As is the case elsewhere in the western world, employment levels in major industries are running down rapidly. This is a consequence of the exhaustion or low-profitability of reserves, as well as the obsolescence or lack of competitiveness of older plants. The failure of the communist system to generate and recycle profits often exacerbated the problem as there was no or little investment in new plants or new industries, especially in countries such as the Czech Republic. Dereliction is concentrated in areas around the original mineral deposits, rather than being scattered evenly across the country. Affected areas were often unable to diversify as the industries already in place employed all available labour at their peak time. Now economic growth is negative or insufficient, leading to high unemployment, and forcing the authorities to try to attract external investment. However efforts are hampered by the atmosphere of obsolescence and neglect caused by dereliction.

Point two: compared to other countries, the Czech Republic has a small number of very large uranium mining sites (for example in Ralsko, in the district of Česká Lípa). A few well-known sites probably merit separate investigation and assessment because of their special characteristics, their large scale and negative perceptions worldwide.

Point three: like other modern states, the Czech Republic has a number of major military bases and other defence installations that have become redundant as part of the "Peace Dividend" following the collapse of the communist block. Formerly incorporated as part of the Austro-Hungarian Empire and more recently forming the western frontier of the Soviet military might, the Czech Republic has a relatively dense network of such sites looking for new uses. These vary enormously. Many are at some distance from large cities. Most, but not all, contain major structures – both above and below ground; and quite a few are contaminated. Official secrecy about such facilities means that the extent of damage and contamination is unclear, while the former use and future potential of the land and buildings are difficult to determine and will often need detailed investigation. The sites are mostly in the hands of the State. Occasionally the authorities have seemed overly keen to pass on problems to new private sector owners, even offering sites at nominal or zero cost in order to make ownership attractive. However, even with such special deals, the costs of transforming such land remain prohibitively high. Thus re-use is rendered uneconomic and the land lies fallow.

Point four: the Czech Republic's long established and diverse industrial base has a wide range of more specialised manufacturing sites (particularly in the field of chemicals, power generation, metal fabrication and precision engineering) that are already redundant or soon will be. Industry must replenish itself and up-date its technology and techniques, thereby improving competitiveness. This is normally an on-going and piecemeal process, occasionally involving major closures or structural adjustments that require special efforts. The failure of communist planners to focus less on heavy industry, and instead increase the production of consumer goods as well as investing in efficient and competitive equipment, has tended to worsen the problem in the Czech Republic (and elsewhere). The end-result has been the speedy dereliction of many large factories in recent years, with more likely to follow in the near future. Such sites are located across the country, with pockets concentrated in the larger manufacturing centres, such as Plzeň and Brno. In smaller settlements the problem can be especially acute, most notably in those towns dependent upon one industry or company. Sometimes the sites are hidden; sometimes they are prominent. They can be particular blights on the landscape or may not leave any visual scars. Sometimes they constitute a health hazard, sometimes not. They usually contain major structures and are expensive to transform to a sufficiently good state for re-use. Although there are large derelict factories in small villages, most of them are found in urban areas. Some will be suitable for re-use, while others are poorly located for modern industry. In such areas a mixed use, with housing or amenity space, may be more appropriate than as sites solely for industry or commerce.

Large basic industries and businesses tend to collapse sequentially, with the weakest first and the strongest last; however the demise of any particular factory or mine does not necessarily bring with it an instant cessation of all activities on that site. The replacement of those industries and associated jobs provides particularly acute challenges in terms of timing, adaptation and location. This means that a successful local economic development plan must integrate the advanced planning of new sites with the restoration of old ones.

2.2 Difficulties in Quantifying the Situation

There appears to be no single source of data on damaged land in the Czech Republic, with various Ministries, public bodies (e.g. The Fund for National Property), municipal authorities and some large industrial enterprises each providing a piece of the picture. Clearly these various sources must be combined to create a single comprehensive database of all sites, sub-divided into different categories. This must be kept up-to-date in order to track the continuing restructuring of the Czech economy – new sites are likely to appear for many years to come. In the meantime it is enough to recognise that there are several hundred (if not thousands) of such sites scattered across the country, with particularly dense concentration in larger towns and cities, as well as in peripheral mining areas, both active and abandoned.

An estimation of the true scale of reclamation needed in the Czech Republic and of the costs of a comprehensive land reclamation programme must await a full survey. But the areas involved will cover thousands of hectares and the final costs of returning sites to economic use can be expected to be substantial. A very small number of large, highly contaminated sites, such as those associated with uranium mining or chemical production, may be very expensive to reclaim, and form potential hazards to neighbouring countries (via, for example, river-borne contamination following an industrial accident). These areas should be seen as a common "European" problem – part of the Soviet industrial legacy. It should then be possible to apply to the European Union for financial assistance to help with land recovery.

Czech cities are facing considerable expenditures in efforts to improve their physical fabric, such as the restoration of buildings, the lighting of buildings and other features, improvement of pavements, creation of traffic-free areas and resurfacing of footpaths. Such concerns, as well as investment in the country's heritage and its environment, are both necessary and praiseworthy, and should be encouraged whenever resources are sufficient. This commendable policy should be extended to removing the scars of the industrial heritage, especially where they affect local people's quality of life, lead to the under-use of land assets and undermine the potential benefits of investment in urban areas.

3 Regions Undergoing Structural Change

According to the regional policy of the Czech government introduced in the year 2000, there are two main regions undergoing structural change on EU NUTS 2 level[2]: the North-West region (composed of the regions Ústí and Karlovy Vary) and the Moravia-Silesia Region[3]. These are defined as problem regions (see map). In 2000 the Czech Government started several development programmes to aid these problem regions. The programmes are:

- Regional Development Programme of North-West Bohemia
- Global plan for the revitalization of the coal field basin in North-West Bohemia
- Regional Development Programme of the Moravia-Silesia Region.

On level NUTS 4 eight declared districts within the two bigger problem regions are undergoing difficult structural change. These are the districts Most, Teplice, Chomutov, Děčín and Louny in North-West Bohemia and the districts Ostrava-City, Frýdek-Místek and Karviná in Ostrava Region. The districts undergoing structural change were defined according to the following criteria:

- the ratio of employees in the industrial sector to those in non-industrial sectors;
- the development of employment in the industrial sector since 1990;
- unemployment characteristics;
- the number of entrepreneurs per 1,000 inhabitants.

The five mentioned districts in North-West Bohemia have in total about 600,000 inhabitants. The region has rich mineral resources, especially lignite coalfields, as well as power and chemical industries connected with the coal fields (www.czech.cz). Economic activities in these fields had, and still have, a negative influence on the environment.

---

2 In the period of January 2000 to December 2002 the territorial structure of the Czech Republic changed. From a political-administrative point of view there are now two subnational levels: the 14 regions (kraje, NUTS 3) and the communities (obec, NUTS 5). For statistical reasons and for reasons of regional policy there are also 77 former districts (okres, NUTS 4) and 8 NUTS 2 regions (cf. Červený/Andrle 2000).

3 Moravia-Silesia Region is both a NUTS 2 and NUTS 3 region (cf. Červený/Andrle 2000).

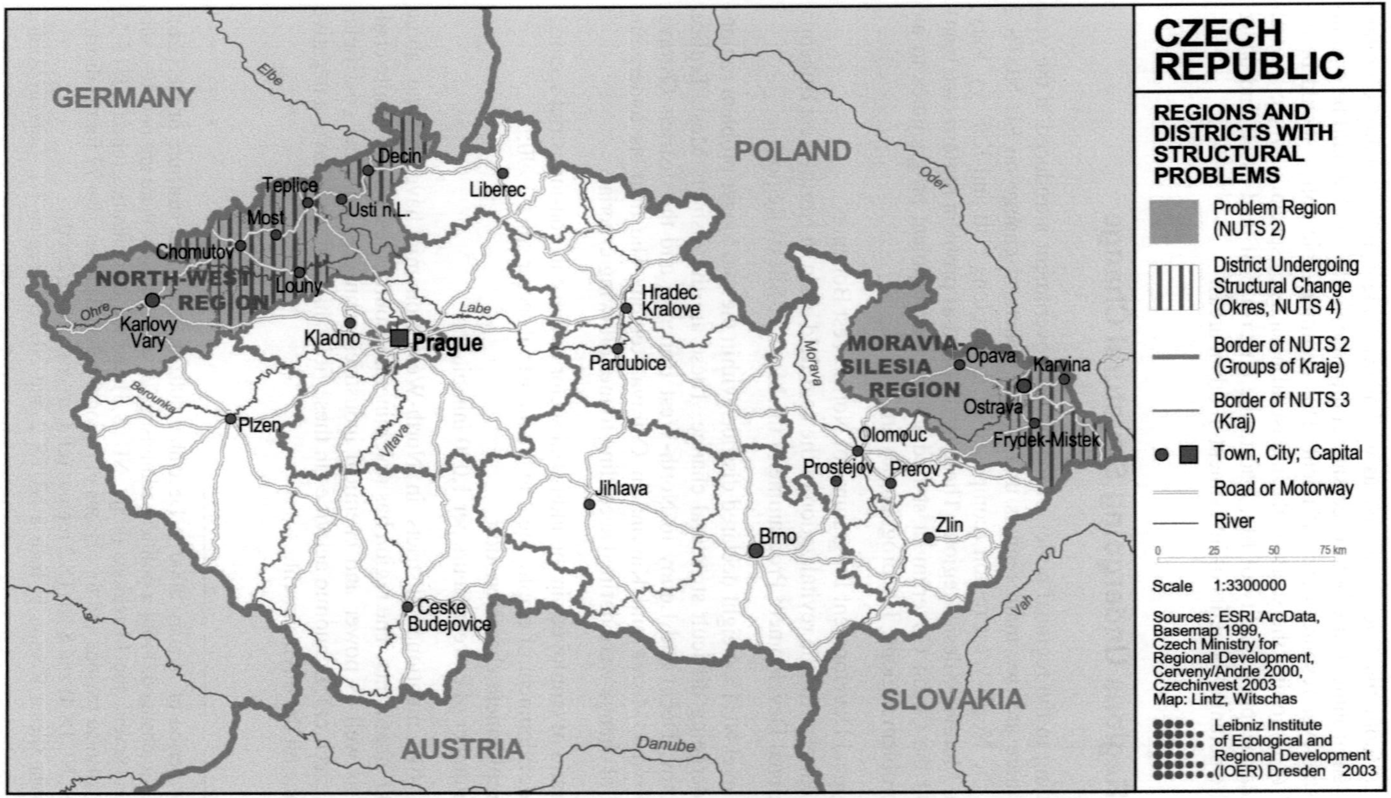

Map: Czech Republic – Regions and (former) districts with structural problems, administrative structure in 2002

Table: Districts undergoing structural change in North-West Bohemia and the Moravia-Silesia Region

Districts	ČR total	Most	Teplice	Chomutov	Děčín	Louny	Ostrava City	Frýdek - Místek	Karviná
Area	-	467 km^2	469 km^2	935 km^2	909 km^2	1,118 km^2	214 km^2	1,273 km^2	347 km^2
Number of inhabitants (31.12.1999)	-	119,178	129,909	125,427	133,768	86,094	321,263	228,291	283,128
Density of inhabitants per km^2	130	256	276	134	147	77	1505	180	819
Rate of employees in industry and building in % (end 1999)	45.2	57.2	52.4	51.0	50.7	44.4	51.4	59.9	54.4
Entrepreneurs per 1000 inhabitants	-	122	124	129	73	131	347	127	108
Rate of unemployment in % (31.3.2000)	9.47	20.59	16.39	17.94	14.04	17.57	16.71	15.07	18.80
Main economic sectors (in terms of GDP)	-	industry	industry	industry	industry, tourism	industry, agriculture	industry	industry, tourism	industry
Main branches of industry	-	mining, chemical industry, power generation, mechanical engineering	mining, mechanical engineering, power generation, glassware industry	mining, mechanical engineering, power generation	mechanical engineering, textile industry	mechanical engineering, ceramic industry, food industry	metallurgy, mechanical engineering, chemical industry, power generation, food ind.	metallurgy, mechanical engineering, chemical industry	mining, metallurgy, mechanical engineering, chemical industry, power generation

Source: Czech Statistical Office, 2000 and 2001

Structural change during the past few years has caused job losses in the industrial sector, and the service industry has expanded too slowly to compensate, resulting in an increased rate of unemployment varying between 14 % in Děčín and almost 21 % in Most (2000). The share of the workforce in the region employed in industry and building ranges between 44 % in Louny and 57 % in Most (see Table).

Industry is also the main employer in the Ostrava region. Karviná, Frýdek-Místek and Ostrava City have together a population of about 830,000 inhabitants; in 2000 the percentage of the workforce employed in industry and building ranged between 51 % in Ostrava City and 60 % in Frýdek-Místek. The unemployment rate varies between 15 % in Frýdek-Místek and 18 % to 19 % in Karviná (see Table). In the 1980s the region showed the highest gross industrial output in the Czech Republic.

The prevailing industries are coal mining, metallurgy and power generation, with a significant concentration on heavy engineering. Some other sectors can also be found, e.g. chemical and pharmaceutical industries, textile industry and food processing. Many older industrial enterprises are in a critical economic situation. Currently a restructuring process is taking place in coal mining and metallurgy. Since 1999 a revitalisation agency has been given the job of restructuring industrial companies and preparing their future sale to strategic partners.

4 Recommendations

4.1 Ten Guiding Principles and Key Issues

The following guiding principles should underlie the creation of a Czech National Land Reclamation Programme:

1. Actions should be additional, and help to accelerate the pace of development (public money should not be used to subsidise projects that already have sufficient financing).
2. Land reclamation measures should not relieve companies and public organisations of their past, current, and future legal obligations regarding the limiting of pollution, contamination, dereliction and visual blight.
3. Dereliction must be seen as a failure of market forces, and public expenditure accepted as the only way to tackle the problem.
4. Land reclamation should be viewed primarily as a contribution to economic development rather than as a measure directed towards environmental recovery or pollution/contaminant removal. Improving the environment is an economic factor in so far as it removes a potential barrier to investment.
5. The main purpose of the programme should therefore be the restoration of land for economic use and development. Industrial and commercial land has a high priority, with the removal of serious health hazards being top priority for immediate action.

6. All available sources of funding and expertise should be sought, with good co-ordination between central government departments and local authorities.
7. A new central unit, under government control, will set priorities and standards, with on-site work being supervised by local authorities.
8. Design and construction work should normally be undertaken by sub-contracted companies.
9. Land should generally be brought into public ownership for reclamation.
10. Every effort should be made to demonstrate the beneficial effects of reclamation at international, national and local levels, and to build a political consensus in the Czech Republic for the creation and maintenance of a national land reclamation programme to be completed within 50 years.

To implement these guiding principles clear definitions of the following terms will be needed: dereliction and reclamation, financial resources, the responsible ministry for process management, and the work, service and management costs that are eligible for funding (this includes procurement and validation of delivery). Other terms such as grant structure, selection of processes, funding commitments (time, circumstances) and authorisation of payments have to be clearly understood, as well as details regarding the return of income to the funding body and assurances that government's legitimate requirements are satisfied.

4.2 Recommended Approach

It is recommended that the following approach be adopted when addressing principles and issues. The central government has to create a body which:

- prepares the overview, policy and programme; procures and provides funds; maintains the national data-base;
- prioritises bids and creates the national programme; funds land purchase by local authorities prior to reclamation; funds 100 % of reclamation costs of local authority projects and monitors the reclamation; establishes a grant of up to 80-90 % for reclamation by private owners and funds owners directly;
- researches new reclamation techniques; gives guidance on using those techniques and propagates optimal practices; encourages innovation and reviews past projects; ensures the attainment of value for money and undertakes international networking.

Local authorities and municipalities should:

- gather site data; create more detailed local databases; send summarised data to the centre; apply to the centre for help;
- prepare, outline, design and deliver proposals for funding (describing site, location and reasons for reclamation, local property needs/demands, end use, etc.); once approval is given, prepare detailed proposals;
- purchase land for reclamation; implement land reclamation measures using in-house teams or tendered firms (the latter is preferable as local capabilities are

thereby strengthened and most local authorities will find it impossible to employ all the needed specialists);
- submit monitoring forms; pay any income owed to the central unit; allow access to sites.

In the private sector, owners of property should:

- submit a proposal for the reclamation of a derelict site to the central unit; once they receive approval and funding, carry out the work;
- allow the central unit access to inspect the work in progress, in accordance with agreed schedules, and ensure inspection after completion; return monitoring forms and pay any income owed to the central unit.

4.3 Necessary Actions

Legal Matters

The first matter to be dealt with is the creation of a national land reclamation agency. If the Czech government wishes to give land reclamation powers to a new institutional unit then the necessary legislation must be prepared and approved. If it decides to award the new powers to an existing organisation such as CzechInvest (Governmental Foreign Investments Agency), then amendments to that organisation's founding Charter must be drawn up.

The new legislation must define dereliction and land reclamation, set out the role of derelict land reclamation in the country's economic polices, place it in the context of existing environmental protection, pollution, and contamination control legislation, and establish the economic purpose of land reclamation. The role proposed for local authorities in the reclamation process must be given a legal footing. Grant delivery mechanisms and their control and administration must be specified, with (ideally) 100 % grants for projects proposed by local authorities and 80-90 % grants for private owners. Furthermore it must be established how and when the public sector will purchase land for reclamation, how the land should be disposed, and how any income from the reclamation process or disposal will be shared.

Another important issue is the distinction between short-term and long-term funding of land reclamation. Short-term projects have little alternative to direct financing from the national budget. A government ministry will be involved in economic or industrial development, as land reclamation is needed to support local economic development. In the long term it may be possible to finance programmes using environmental taxes (e.g. tax on the transfer of agricultural land to development, on land-fill waste disposal and, even tax on pollution). However such sources of income are unpredictable and may fluctuate highly; therefore they should be seen as a supplement to direct budget lines in the sponsoring ministries' expenditure plans.

The government should also define how land reclamation relates to land use planning and local economic regeneration strategies, and specify the necessary degree of consultation with local authorities, local communities and residents living near individual reclamation schemes. Finally, a sufficient programme budget is needed to establish the proposed national agency and to provide it with funding for the first three years. At the end of this period the long-term needs will be defined and a corresponding budget can be established.

Parallel to this the government should review existing environmental protection, pollution and contamination control legislation, to ensure that it is consistent with new legislation. Furthermore, current powers for compulsory purchases must be examined to determine whether they need strengthening or improvement.

Establishing a Start-Up Budget

As mentioned above, an initial budget should be planned by the Czech Government that is sufficient to establish a unit (or new department of an existing body) and fund it for three years.

In order to make an early start on land reclamation, it is recommended that the European Commission's Phare Programme be asked to provide a two-year Technical Assistance project within the Special Pre-Accession Programme for the Czech Republic. This would supply the new Czech agency with a full-time adviser and short-term experts to assist with specific technical issues, as well as a computer network to construct the national derelict land database and give access to external organisations. Further services include the funding of publicity for the launch of the programme and dedicated unit, the creation of a website, and help with the design of reclamation schemes for five pilot projects. These measures should be of interest to the Phare Programme, as many other countries funded by Phare have similar dereliction problems to the Czech Republic that are recognised to hold back economic regeneration. Alternatively, other aid organisations could be asked to help with specific elements of the above activities, especially with the design of pilot schemes. If such approaches fail the government must come forward with the necessary financing.

Establishing a Central Unit

The establishment of a separate Czech Land Reclamation Unit, or a new unit within CzechInvest or some other national executive agency, should be an immediate priority. The Czech Government, here meaning the ministries responsible for economic and industrial development as well as regional development, should set up a steering group to give advice on the best organisational "location" of the Unit. This includes identifying the "lead" ministry to which the unit will report and from whom it receives funds and its mandate. The steering group should also try to build a wide political consensus for a national land reclamation programme, communicate the benefits to local authorities, and discuss specific issues with those older industries that cause dereliction or which own derelict sites. This steering group will eventually evolve into a supervisory board for the Unit.

References

Arnold L, Griffith G (2000) A Strategy for Industrial Land Reclamation in the Czech Republic. Executive Summary. Presented to CzechInvest, May 2000 (http://www.levego.hu/bzona/gescz.htm)

Canadian Commercial Corporation (1993) Reclamation of Northern Moravia Coal Lands, Calgary

Červený M, Andrle A (2000) Czech Republic. In: Bachtler J, Downes R, Gorzelak, G (2000) Transition, Cohesion and Regional Policy in Central and Eastern Europe, pp 85-97

ERES Ltd, SAURA s.r.o (2000) Study for the Development of Strategic Sites in North Bohemia, Brno

Havrlant J (1999) Negative Influences of Coal Extraction in the Mining Areas of the Karviná Region. In: Moravian Geographical Reports, Vol. 7, No. 1, pp 56-60

Horáková J (2000) The Karolina Site in Ostrava Experience, Ostrava

Kunc J (1999) Transformation of Industrial Production in the District of Brno-Province. In: Moravian Geographical Reports, Vol. 7, No. 1, pp 48-55

Ministry for Regional Development of the Czech Republic (1998) Urban brownfields: A specific situation in the Czech Republic, Ústav územního rozvoje, Brno

Norwest Mine Services (1994) Czech Republic – North Bohemia Brown Coal Mining and Reclamation Study, no place of publication indicated

Germany: Tackling the East-West Divide

Gerd Lintz[1] and Karl Schmude[2]

[1] Institute of Ecological and Regional Development, Dresden
[2] Ministry for Labour, Building and Regional Development Mecklenburg-Vorpommern, Schwerin

1 Introduction

In the last few decades many industrialised cities and regions in Europe have been undergoing a remarkable process of structural change. Some of them have succeeded in adapting to the changing circumstances. However, in many cities and regions the decline of industries such as coal mining, steel production, textile production and shipbuilding has not been compensated by a growth in new sustainable industries. These cities and regions, often termed "old industrialised", are suffering high unemployment rates, decreasing incomes, deformed urban structures and ecological damage. The challenges for industrial cities and regions with structural problems are of a highly complex nature.

Structural change in Western countries has been driven mainly by the market economy, and changes have been relatively slow and continuous, if at times painful. In the socialist Central and Eastern European countries the moderate pace of structural change as driven by central planning was abruptly superseded by an unprecedented transformation from socialism to capitalism. This necessitated drastic and wide-ranging adaption measures, if these countries wished to integrate into the structures of the world economy.

Germany is situated in the heart of Europe. With around 82 million inhabitants it accounts for almost one fourth of the gross domestic product of the European Union (EU). A recent history of division and reunification has given Germany experience of the problems of sustaining structural change in a market economy as well as a system transformation from socialism to capitalism. Drawing from existing studies, the aim of this paper is to describe and compare the problems of cities and regions in western and eastern Germany while examining the strategies adopted there to tackle these problems.

The article is structured as follows: Following the introduction, the historical development and current situation of old industrialised cities and regions is outlined, stressing the quite different paths of development in western and eastern Germany (second section). The third section deals with the strategies adopted by the five political-administrative levels in today's Germany to overcome specific problems. In the last section some conclusions and future perspectives are given.

2 Two Different Paths of Development

German cities and regions have shown a diversity of development in the last 60 years. This is primarily due to the entirely different trends of general development in western and eastern Germany (section 2.1). A glance at the map shows that an interesting geographical pattern of old industrialised cities and regions has emerged (section 2.2). The special problems and the reason of decline of such cities and regions are described in section 2.3.

2.1 Trends of General Development

2.1.1 Economic Trends

After World War II, the basic conditions of urban and regional development on either side of the newly imposed internal border of divided Germany could not be more different. The paths of development of West Germany, consisting of the eleven old states (alte Bundesländer), and East Germany, consisting of the five new states (neue Bundesländer)[1], led in diverging directions. After the decision was taken in favour of an open social market economy, the US Marshall Plan fostered the economic rebuilding and development of *West Germany* to continue with its pre-war dynamism. Rapidly increasing wealth for the great majority of the population and unemployment rates of less than 2 % characterised the resulting "Economic Miracle".

At the end of the 1960s and the beginning of the 1970s (first oil crisis) growth rates began to decline. Aggravated by an expanding population, the unemployment rate increased steadily to an average level of around 10 %. This development was accompanied by ongoing structural change: the share of the national workforce employed in industry decreased, while the tertiary sector saw expansion (cf. Ambrosius 1994, Schröter 2000).

[1] Following reunification the difficulty arose of how to name the territory which constituted the Federal Republic of Germany (FRG) and the German Democratic Republic (GDR), often referred to as West and East Germany in political terms. The (old) FRG consisted of eleven states, now called the "old states" (alte Bundesländer). The GDR dissolved into five "new states" ("neue Länder") and acceded to the FRG. West Berlin merged with East Berlin. Many years after reunification it is becoming less and less appropriate to use the designation "old" and "new" states. So "western" and "eastern" Germany are being increasingly used in the geographical sense (although the German language does not distinguish between west and western or east and eastern). As a consequence, Berlin, one of the "old" states which belonged to West Germany, is now more frequently viewed as part of eastern Germany.

With increasing integration into a dynamic world economy, competition became more fierce for domestic producers. Branches which manufactured products at a relatively low technical and labour-intensive level were particularly affected. For example, the coal-mining industry, textiles and the steel industry lost market share to low-wage countries. In addition, demand for certain products sank for purely technological reasons, affecting the coal and steel industries.

Restricted by high wage levels, not all affected enterprises and factories were able to boost competitiveness through product innovations or rationalisation. Instead many were forced to close. Or, in the case of a successful lowering of labour costs by way of rationalisation, only a reduced workforce could be retained. The spatial concentration of those branches affected by the crisis and certain unfavourable local and regional factors resulted in a spatial concentration of problems such as unemployment and low financial power in some cities and regions.

Reunification in 1990 was initially followed by a short economic upswing in western Germany before economic growth was stifled by the high ensuing costs. Existing structural problems such as an overly expensive social security system and huge public debt were exacerbated. The table below shows that employment in the period of 1990-1998 decreased by 1.2 %, with the unemployment rate reaching almost 10 % in 1998.

Table: Employment and unemployment in western and eastern Germany

	Employees liable for social security contributions		Labour-force participation rate		Unemployment rate
	Development 1990-98	% of inhabitants 1998	1989	1998	1998
Western Germany	-1.2 %	33.0 %	50 %	48.1 %	9.8 %
Eastern Germany	-36.8 %	33.6 %	75 %	53.3 %	18.2 %

Source: BBR 1999a; Employment rate 1989: BMBau 1991, Employment rate 1998: Statistisches Bundesamt 1999.
Labour-force = Gainfully employed persons + unemployed persons; Labour-force participation rate = ratio of labour-force to total working age population; Unemployment rate = proportion of the labour-force who are unemployed.

After World War II, *East Germany* was obliged to pay reparations to the Soviet Union for war damages. The new socialist state was integrated into the Council of Mutual Economic Aid (COMECON) of socialist countries in Europe[2]. Despite the waves of highly selective East-West migration before the erection of the "iron curtain" in 1961, a strong tradition of industrial production gave the German Democratic Republic, GDR, (Deutsche Demokratische Republik, DDR) a leading position in the Eastern bloc. However, economic development rapidly fell behind that achieved in West Germany, with investment in physical capital being particularly weak. This led to low labour productivity, low quality products and high consumption of resources and environment. The GDR also experienced moderate

[2] Until 1989, more than 70 % of exports were shipped to COMECON countries, 40 % went to the former USSR (Hohlfeld 1999).

structural change, resulting in inefficient structures, i.e. excessive industrialisation. This was deliberately driven by central planning and did not cause unemployment.

Reunification brought an unprecedented transformation from socialism to capitalism. Less than one year separated the falling of the wall in November 1989 and the achievement of national unity in October 1990. The GDR dissolved into five federal states and acceded to the Federal Republic of Germany, FRG, (Bundesrepublik Deutschland, BRD). East and West Berlin merged. The process and outcome of these dramatic changes has been extensively documented (e.g. BBR 2000b and 2001, DIW 1999 and 2000, DIW et al. 2002, Härtel 2001, IWH 2001, Mayr and Taubmann 2000, Pohl 2000 and Sinn 2002).

While in western Germany only a few branches of industry underwent contraction over a lengthy period of several decades, in eastern Germany all branches were negatively affected in a historically unprecedented way within an extremely short period of time (Pohl 1995). The pressure to adapt was tremendous. Industrial employment fell drastically, from approx. 3.5 million people in the year before reunification to just under 730,000 in August 1993. In mid-1993, eastern German industry contributed only 3-5 % to the overall German industrial production (Fischer 1994) – this can be considered a "structural break" and a process of "de-industrialisation" (for example acc. to Scholz 1994, 9). Eastern Germany's population decreased from 16.6 million in 1989 to 15.2 million in 1999 (Statistisches Bundesamt 2000), mostly a result of net migration to western Germany.

With reunification, the complete economic, legal and social welfare system of the FRG was applied to the former GDR (including introduction of the West's Deutschmark). Subsequent analysis reveals that the dramatic economic slump had four causes. Firstly, the currency conversion of the Mark (East) into the Deutschmark (DM, West), with wages and salaries fixed at a 1:1 rate, was problematic (Mulke 2000). Since an improvement in the low productivity of the eastern German economy to compensate the increased wages was impossible factories were hard pushed to operate at a profit. Secondly, conversion of the debts of former GDR-enterprises was, in hindsight, unwise. Loans in the GDR had an entirely different function than in countries with a free market economy: enterprises were neither able to independently raise loans nor to use loans for investments. This fact was ignored and enterprises continued to be burdened with the converted loans after 1990.

Thirdly, the way property relations were handled turned out to be an impediment to investments. It was seen as of utmost importance to return unlawfully expropriated manufacturing facilities and other real estates to the lawful owners, even when unclear property relations led to considerable delays in investment. A fourth negative aspect was that demand for domestic products collapsed in favour of goods produced in western Germany and western Europe. Demand from previously predominant markets in other socialist countries also dropped, partly due to the fact that goods had to be paid in DM and partly because of economic problems in those transforming economies. The result was not, as hoped, a second "Economic Miracle", but instead a painful and still ongoing structural breakdown.

The rapid drop in industrial employment after 1993/1994 finally bottomed out, to be followed by an increase in production, resulting from the implementation of strong economic support. From 132.5 industrial employees per 1,000 inhabitants in January 1991, the figures for eastern Germany plunged within three years to 46.5; in June 1997, they had risen somewhat to 73.0, still far below the western German average of slightly more than 100 (Nolte and Ziegler 1994; Statistical Offices of the Federal Government and the Federal States). Even allowing for over-estimates in the GDR's industrial employment (many non-industrial services were provided by companies) this development is quite significant.

In the period 1990 to 1998 the number of employees liable for social security contributions (including industrial workers) fell by a total of more than one third (cf. Table). It should also be mentioned that the high unemployment figure in comparison to western Germany is partly to be explained in the much higher labour-force participation rate. The unprecedented number of redundancies was cushioned by numerous social, labour market and economic-political measures. This is one reason why, with 2.36 million jobs lost in eastern Germany by 1998, only 1.38 million people were reported unemployed (DIW 1999). The 18 % unemployment rate was almost twice as high as in the western part of the country. In eastern Germany the gross monthly income in the manufacturing industry is only 75 % of that in western Germany. However, overall economic productivity is still only around 60 % of the western level (DIW 1999)[3].

Of great importance for the economic situation in eastern Germany is the privatisation of around 8,000 formerly state-owned enterprises. By 1994, almost all enterprises, parts of enterprises, factories and other property were free for privatisation, reprivatisation, communalisation or liquidation (BVS 1995). It should be noted that most enterprises and factories were bought by West German or foreign companies. This distribution of property and control rights might become a problem for eastern Germany in the long term. However, other concrete difficulties connected with branch plants (lack of headquarters, poor levels of research and development) will probably play a more important role than the outflow of investment income.

Serious problems could also arise from the wave of young and qualified employees migrating to western Germany as a result of the economic situation. Together with the dramatic decrease in births – insecurity led to a collapse of the total fertility rate from 1.52 in 1990 to a world record low of 0.77 in 1994 (Schaich and Fleischhacker 2000) – demographic developments can be expected to greatly burden the economy and public budgets in the future (cf. BBR 2000a, 9-16).

After the initial slump of the eastern German economy, GDP then picked up, growing at a much higher rate than in western Germany. Although the "Aufbau Ost" (the attempt to adjust eastern Germany to western standards) has made clear strides, the process of catch-up stopped in 1997, with stalled and falling growth rates (SVR 2002, 178). A considerable prop to improved living standards is still financial aid from western Germany. Fortunately industry is steadily growing, although positive effects are still countered somewhat by the decline of the con-

[3] The latest study (DIW et al. 2003, 9) says that eastern Germany in 1998 showed 67 % and in 2002 showed 71 % of the productivity of western Germany.

struction sector; this sector previously gained an excessive share in the economy in the first years after reunification.

2.1.2 Environmental Trends

Air, water and soil have been heavily polluted in the industrial regions of both *Western and Eastern Germany*. Furthermore, the landscape of almost all industrialised regions has been transformed by specialised infrastructure such as freight train railways, pipelines and high-tension power lines, often appearing overdimensioned and unsuited for later reutilisation (recreation, housing, commerce).

The strong spatial concentration of the heavily polluting coal, steel and chemicals industries has produced regional centres with serious environmental pollution. Although the large number of pollutants makes generalisations difficult, the most important sources of environmental pollution are the following:

- Combustion processes in power plants and heavy industry, causing massive and far-reaching air pollution with sulphur oxides, nitrogen oxides and dust.
- The air quality in regions with coal and steel industry was also seriously impaired by hydrocarbons from coking plants and exhaust gases from blast furnaces. The hydrocarbons also caused deep-reaching soil pollution.
- The chemical and textile industries emitted a large number of hazardous substances into water, air and soil. Heavy metals and some persistent organic byproducts still constitute a problem decades after production has ceased.

The situation in the GDR was worsened by the fact that the entire energy and raw material industry was largely based on technical processes from pre-war times. The use of lignite as the only energy source and a hazardous system of waste treatment produced high emissions of pollutants. Metallurgy industries in the GDR were forced to use inferior ores, requiring a great deal of energy for extraction and processing (SRU 1995, Tz. 532-561).

In West Germany modern environmental legislation was introduced at the beginning of the 1970s, based on the polluter-pays principle[4]; economic structural change then led to a continuous improvement in water and air quality. In eastern Germany air and water quality improved greatly after reunification. The rapid structural change and implementation of West German environmental standards forced polluting industries to modernise or face closure.

Notwithstanding the unquestionable improvements in air and water quality in west and east, the remaining problems of soil pollution on old industrial sites[5], of

---

[4] The "polluter-pays principle" can be regarded as the cornerstone of most environmental legislation. It states that the cost for environmental damage (previously borne by the public purse) has to be paid by the polluter – these external costs have to be internalised.

[5] According to Hentrich et al (2001), of the almost 300,000 suspicious sites in Germany (Dec 1998), only 10-20 % will eventually have to be cleaned up. The remaining sites will either be classified as harmless, put to new use or closed for public access.

devastated landscapes and of groundwater deficits in open-cast mining regions[6] still have to be dealt with. Environmental reclamation is, however, costly and time-consuming (UBA 2000a). One special case requiring attention is the radiological contamination of uranium ore mining areas in the southern parts of Thuringia and Saxony.

The fact that many companies have folded in the course of structural change nullifies the "polluter pays" principle, passing the task of soil remediation to the buyer of a contaminated site or to public authorities (who are often not able or willing to raise the required funds). Costs can be quite considerable; according to the planning ministry of Northrhine-Westphalia (MSWKS 2001), the remediation costs for the 110 ha. site "Phoenix-West" in Dortmund – a former blast furnace – will be more than 36 million euros.

The economic ramifications are pointed out by Henckel (1986, 66): the high real property prices in economically attractive regions mean that "wastelands practically do not constitute a problem", as private enterprises are prepared to buy and use these areas. However in "old industrialised or peripheral regions which have lost their attractiveness for new types of industries (...) the old contaminations from the past – in the broadest sense[7] – must often be removed, involving high public expenditure." Several years may go by after a factory closes before the area becomes available for reutilisation (Tettinger and Mann 1994; SRU 1995, Tz. 260). The reutilisation of brownfields thus becomes quite difficult.

It should be emphasised that unlike other problems which old industrial regions are experiencing, environmental damage is not due to the economic breakdown; it is rather the immediate consequence of former economic activities. The issue seems to receive more attention when a given industrial region is undergoing serious decline; when industry is booming, bad news about negative impacts tends to be dismissed, for obvious reasons.

Today, we have the situation that air and water pollution has actually been decreasing for years, and great success has been achieved in soil remediation. However the persistent and unjustly negative environmental image of western and eastern German industrial regions is a key factor in discouraging external investment.

---

6 One of the largest open-cast mining areas in Europe is Lower Lusatia (Niederlausitz) where approx. 800 km^2 of land are devastated, partly due to the political decision of basing the GDR's energy supply on domestic lignite. Excessive pumping of water to keep pits dry has lowered the groundwater table by as much as 80 m on approx. 2,500 km^2, with a considerable impact on agricultural, forestry utilisation and biotopic conditions. Now that mines are closed, large amounts of water are required to recover water deficits; the amounts required and necessary quality mean that this cannot be left to natural groundwater refill. Instead water must be supplied via pipelines over long distances (Georgi 1994, 350).

7 This includes buildings and specialised infrastructure which characterise these areas.

2.2 Distribution of Old Industrialised Cities and Regions

Influenced by various factors of structural change and the different powers of adaption shown by individual areas, a certain pattern of spatial distribution of old industrialised cities and regions has emerged. As the map shows, cities and regions suffering from incomplete structural change are only thinly scattered across the eleven old states[8] with much more extensive problem-areas in the five new states (encompassing Mecklenburg-West Pomerania, Brandenburg, Saxony-Anhalt, Thuringia and Saxony). As the situation in unified Germany is heterogeneous, the delimitation of industrial cities and regions requires a pragmatic approach.

The map shows the areas in western Germany given objective 2 status in EU regional policy (declining industrial areas, 1989-1999). With its comparatively low economic power, the whole of eastern Germany is eligible for objective 1 status (regions lagging behind in development). Since all cities and regions which were industrialised before reunification have experienced with dramatic structural change, they can now be classified as old industrialised areas, or as industrial areas with structural problems. The map shows these cities and regions, distinguishing between industrialised and strongly industrialised areas (data from 1971, cf. Scherf and Scholz 1990).

In *western Germany*, regions with the following previously dominant industries were particularly affected: hard-coal mining, iron and steel production, ship building, textile and clothing industry. The result was transient local unemployment rates of up to 20 %, as for example in the Ruhr area (Danielzyk 1992).

The densely populated and strongly urbanised Ruhr District in the centre of North Rhine-Westphalia (pop. five million) is by far the most important old industrialised region in western Germany – and at one time the leading region in Europe (cf. e.g. Blotevogel 1997, Bömer 2000, Dürr and Gramke 1993, Hamm and Wienert 1990, Schrader 1998 and Wood 1997).

Economic development and a changing employment structure in this conurbation has been characterised by a more or less continuous decline in industrial employment since the 1960s and 1979s. Hard-coal mining and steel production have been worst hit, with 500,000 jobs being lost in these two sectors alone (Wienert 2000, 28). Since there has not been a compensatory increase in jobs in other industry branches or in the tertiary sector, unemployment rates have also risen. Development has varied from city to city, but all are confronted with above-average unemployment rates. The Saarland, Germany's second (much smaller) old coal and steel region, is suffering similar problems to those faced by the Ruhr district.

The shipbuilding and harbour industry was also forced to make closures and reduce capacity, causing long-term problems in some port-cities, e.g. Bremen, Bremerhaven and Kiel. The significant loss of jobs in the leather, textile and clothing industries has badly affected regions such as the sparsely populated West-Münsterland Region in the southwest of Lower Saxony and the area around the north Bavarian textile city of Hof.

[8] Including the western parts of Berlin, formerly an exclave in East Germany.

While unemployment did not officially exist in the GDR, after 1989 the unemployment rate of *eastern Germany* rose to unprecedented heights, in some places exceeding 25 %. In line with the spatial distribution of industry, Saxony, Thuringia and the south of Saxony-Anhalt were particularly affected, as well as some cities and regions in Brandenburg and Mecklenburg-West Pomerania (cf. Bönisch et al. 1982, Häußermann 1992 and Schmidt 1995).

The following problem areas are specifically mentioned in view of their concentrations of declining branches. The Leipzig Lowlands (cf. Oelke 1998) and the area around Cottbus (Lower Lusatia, cf. Kowalke 1998) are former lignite mining, lignite processing and energy production centres. Some derelict centres of chemical industry can also be found scattered around the lignite deposits, for example, within the triangle of Halle-Leipzig-Dessau.

Another important branch of industry was uranium-ore mining in the south of Thuringia (e.g. Ronneburg) and in the southwest of Saxony (e.g. Johanngeorgenstadt), developed as a joint Soviet-German military venture under great secrecy and with no concept for subsequent remediation. The steel industry, later suffering extensive closures, was concentrated primarily in the cities of Brandenburg (state of Brandenburg), Eisenhüttenstadt (Brandenburg) and Riesa (Saxony). Shipbuilding and harbour activities were located in northern Mecklenburg-West Pomerania; these branches have also seen considerable job losses. The textile and clothing industry has disappeared almost completely, reduced to only a few production facilities, predominantly in western Saxony (Kowalke 1998, 428). Considerable adaptation problems also occurred in heavy machinery construction in Magdeburg (Saxony-Anhalt) and in automobile and machine construction in the Chemnitz region (Saxony).

Some regions have managed to maintain their industrial tradition with the help of modern production facilities. For example optics and precision mechanics is being developed in Jena (Thuringia), microelectronics is flourishing in Dresden (Saxony) and a strong chemical industry can be found the region Leipzig-Halle (Saxony/Saxony-Anhalt). Additional potential growth centres are Berlin, Erfurt in Saxony-Anhalt and Chemnitz in Saxony (SVR 1999).

2.3 Regional Problems and Reasons for Decline

Our general analysis shows that the basic conditions of development in western and eastern Germany have been very different. Weak long-term macroeconomic performance has also complicated the processes of structural change over the whole of Germany. It is not surprising that one can find cities and regions that have been unable to meet the challenges successfully; in each case the concrete difficulties are highly dependent on spatially differentiated factors. The problems of affected cities and regions in eastern and western Germany are in principal similar – with the overall situation in eastern Germany being considerable worse, showing particular areas of weakness.

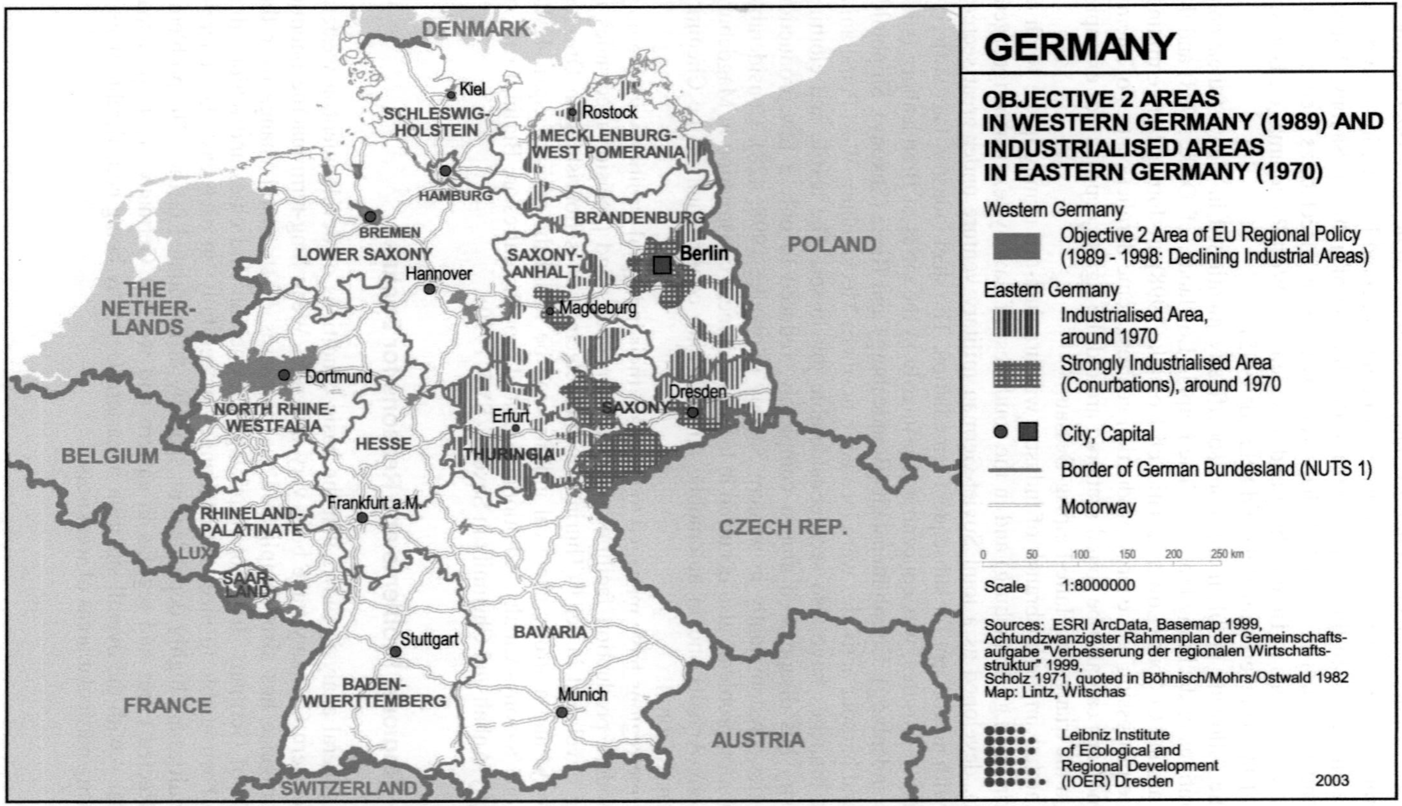

Map: Germany – Old industrialised cities and regions

The heterogeneous distribution of natural resources is reflected in the distribution of negatively affected branches across the country. The higher the share of such branches in the urban or regional economy the bigger the challenge posed by structural change. However, it is important to note that not only the depth of necessary adaptation can vary; the local or regional ability to cope with a certain challenge can also differ (Steiner 1985, 389). The following location factors are decisive for the growth of existing enterprises and the emergence of new enterprises to provide additional employment (cf. Hamm 1991).

(1) The *entrepreneurial potential* of a city or region can influence the pace of adaptation to structural change in a market economy. While one can argue that entrepreneurs and managers in the Ruhr District have concentrated on coal and steel for too long, in East Germany there was almost no entrepreneurial tradition. The principles of the market economy and private-sector management (especially financing and marketing) were largely unknown in the GDR, and have had to be newly learned. This is still a problem today.

(2) In spite of the availability of extensive industrial wasteland, *developed operationally-usable areas* have been lacking. The reasons for this are actual or suspected waste deposits, unclear property relations and dilapidated infrastructure and buildings. In the Ruhr District, the so-called "blockage of land" worsened the situation: old-established enterprises owned the majority of potential areas and refused to sell. A similar effect, at least temporarily, was caused by the insecurities regarding property relations in eastern Germany. Today a lack of areas should no longer play a decisive role.

(3) Traditional industrialised cities and regions usually have a specialised *infrastructure* of ports, railway systems and pipelines, designed to provide bulk goods transportation. This does not meet the requirements of flexible production structures (e.g. just-in-time delivery). It will take many years for eastern Germany to modernise infrastructure in this regard. The federal government and individual states have agreed that the new states will be specially assisted till the year 2019, as part of a financial equalisation policy.

(4) The availability and quality of *labour* is highly important. Often workers have inadequate qualifications or no professional training. Selective migration, such as seen in eastern Germany, can aggravate the situation. With its strong labour unions, the Ruhr District was for a long time a high-wage region, discouraging private enterprise. As already mentioned, labour costs in eastern Germany are too high in view of the lower worker productivity.

(5) The number and size of *educational facilities* is an associated issue. The training of sufficiently qualified personnel is dependent on the capacity of educational institutions. Today, educational possibilities in industrial regions have been expanded from their previous narrow specialisations. Universities, in particular, play an important role in urban and regional development by additionally providing *public research.* The role of research institutions will be addressed in point (8) below.

(6) As hard location factors are becoming widespread, *soft location factors* such as environmental, landscape or urban qualities and the standard of cultural life are gaining in importance. These are especially vital in attracting high quali-

fied employees, because they characterise the image of a town or region and influence managers and workers in creative professions to move to a certain site. While in western Germany a strong environmental policy beginning in the 1970s led to improvements in air and water quality, in eastern Germany the main improvements have come with the rapid industrial decline (Hentrich et al. 2001). With their high spatial concentration of pollution, towns such as Bitterfeld in Saxony-Anhalt or the area south of Leipzig were synonymous with ecological catastrophe.

The urban features of industrial settlements often do not come up to today's standards. Residential areas lack green spaces, housing quality is often poor and the (remaining) industry is a source of pollution. In some cases urban structures were destroyed, as for example in Johanngeorgenstadt (associated with uranium mining) where the historic town centre was deliberately demolished. Cities sometimes formed around mines or factories: Oberhausen in the Ruhr District is only one example of several in that area. In East Germany new industrial cities were planned and built, such as Eisenhüttenstadt in Brandenburg. Industrial wasteland is often situated near the town centre, and needs to be revitalised and integrated into the town. Unfortunately, greenfield development was often given priority over brownfield development (cf. UBA 1998 and 2000). While in the Ruhr District reuse of old areas started relatively early, in eastern Germany there was initially a strong trend of constructing industrial, business and residential areas in the suburban space. Urban deficiencies in East Germany were aggravated by a socialist housing policy which favoured slab constructions, deliberately neglecting pre-war housing in the city centres. Furthermore, the quality of life in old industrialised cities and regions is impaired by cutbacks in social and cultural infrastructure. Despite impressive efforts in the Ruhr area and in large parts of eastern Germany, there are still deficiencies in soft location factors.

(7) The *commitment of local politicians and administrators* also plays an important role in economic development, and the pace of structural change can be particularly affected. In the Ruhr District an attitude which favours the preservation of structures has resulted in delays in implementing sustainable solutions to long-term problems. In eastern Germany a much greater readiness to accept change has been displayed, although unfortunately the necessary experience to ensure success has sometimes been lacking.

(8) Aside from these soft location factors, *"ultra-soft" location factors* have also been identified (Butzin 1996). These are strongly connected to the innovation-oriented co-operation between economy, science and politics. Networks form a basis for co-operation (Fürst and Schubert 1998), with the existence of *research and development* (R&D) facilities in the public and private sector being a core element. Positive attitudes regarding co-operation and innovation can also be helpful in fostering a culture of innovation or a "creative milieu". The milieu in the Ruhr District, for example, is considered to be non-innovative by some authors (Butzin 1993, Grabher 1993 and 1994). Many old networks in eastern Germany have broken down, with new regional networks emerging only slowly (cf. Grabher and Stark 1997). While public R&D spending has grown sufficiently, there is still a lack of R&D effort in the private sector. Since innovation has become increasingly important for urban and regional competitiveness, the exploitation of net-

work and co-operation potentials and "creative milieus" is seen as a key factor in development. Ultra-soft location factors provide a central endogenous potential to activate other material endogenous potentials.

Rising unemployment and stagnant wages in affected industries can trigger further problems if no strong development policy is implemented. These problems can lead to a vicious circle of general decline, weakening the aforementioned location factors. Consumer purchasing power is reduced and rising public spending on social security diminishes the scope of action of municipalities. Expenditures are cut back just as more money is needed to improve location factors for economic revitalisation. Decreasing urban and regional attractiveness often leads to the migration of valuable workers. The town of Hoyerswerda in Saxony provides a prime example of these negative processes. Built as a settlement for an energy and chemical production complex (Hamacher 2001), the number of inhabitants decreased after reunification from 65,000 to 45,000, constituting a veritable urban disaster. Areas in eastern Germany with high migration rates have large numbers of vacant flats and residential buildings, as well as showing general signs of decay and social segregation. These factors tarnish the image of afflicted cities and regions, and eventually inhabitants can lose both their confidence and the will to promote a positive development.

The complexity of the problems facing old industrialised cities and regions demands an integrated approach of different policies and measures at local and regional levels, with crucial elements being the commitment and expertise of local and regional protagonists. Their participation in the formulation of development objectives and the granting of financial aid increases the motivation of all parties and raises the likelihood that measures will be accepted.

All things considered, the situation of old industrialised cities and regions has improved, especially in eastern Germany. The implementation of redevelopment strategies has radically changed the face of many cities and regions, with some formerly depressed areas now flourishing. However these success stories must be weighed against those areas still suffering persistently high unemployment.

3 Interdependent Strategies on Different Levels

As discusses above, a considerable number of cities and regions are old industrialised, especially in North Rhine-Westphalia, Saxony-Anhalt, Thuringia and Saxony. Since the end of the 1960s strategies have been developed and implemented to tackle the problems of structural change. The strategies evolved over decades and varied according to the development of problems, practical experience, specific conceptual ideas and the particular institutional framework. The most influential factors affecting structural change are general economic development, the continuous processes of EU integration and, of course, the reunification of Germany. The latter, in particular, dramatically changed priorities in the resolution of spatial problems. After reunification, attention and financial means were very much shifted to eastern Germany.

Germany is especially marked by its multilevel governance. There are no less than five important political-administrative tiers with the responsibility of formulating or pursuing development strategies for old industrialised cities and regions: the European, federal, state, regional and municipal levels. They have different degrees of authority and a diversity of aims, so that a complex system of interweaved and highly interdependent strategies has emerged. Before turning to the strategies of different political-administrative levels some initial points are addressed.

3.1 Initial Points

Germany has made outstanding efforts to support the development of old industrialised cities and regions, both directly and indirectly. There is a relatively strong orientation towards a "social" or "welfare" state, incorporating spatial specifications. Article 72 in Germany's constitution (Grundgesetz) introduces the aim of establishing "equal living conditions throughout the federal territory" (Tschentscher 2002). Accordingly, the Federal Regional Planning Act (Raumordnungsgesetz) states in paragraph 1: "similar standards of living shall be established in all regions" (Federal Office for Building and Regional Planning 2001, 61).

Strategies to support old industrialised cities and regions are concerned with two aspects. One regards the choice of development aims and supporting measures, e.g. investment incentives for companies, upgrading of infrastructure or environmental remediation (content, policy). Another aspect is the social and institutional process of defining aims and selecting/implementing measures in terms of, for example, the complementary roles of cities and regions, leading to the distinction of top-down or bottom-up approaches (Politics, Polity).

Concerning the first aspect of content, at least three or four strategies to ease structural change can be distinguished. Firstly, existing endangered branches can be safeguarded (i.e. given "economic protection"). This means that companies hit by crisis are stabilised by special support. However in the long run it is often impossible to save old structures and to prevent structural change in this way.

Secondly, new jobs can be created in other industries, particularly in industries which are seen as "future branches", such as microelectronics or biotechnology ("new jobs", "innovation"). This approach stresses the necessity of facilitating structural change, enhancing the framework conditions in general, and targeting new branches. Enterprises must be kept competitive and help given in the founding of new enterprises, either directly (subsidy payment) or indirectly (improved infrastructure etc.).

Thirdly, socially-compatible solutions are necessary for the newly unemployed who have little chances on the labour market because of, for example, age discrimination (early retirement schemes etc., "social cushioning").

Beyond these direct economic objectives, a fourth focus is to make improvements in the environmental situation and the general quality of life. These are major deficits in most industrialised cities and regions ("environmental improvement"). Although all approaches are currently adopted in Germany, the second and the fourth are gaining in importance.

3.2 European Union, Federal Government and States

The federal character of the German constitution influences the way in which structural change is handled. In addition to the Federal Government (Bund) and the 16 states (Bundesländer, Länder) there is a third constitutional level of administration – the around 14,000 self-governing communities. Another relevant issue of increasing importance is the integration of Germany into the European Union. Both the German institutional system and the planning system in general are well documented (in English) by Albers (1999), BBR (2000c) and Turowski (2002).

Strategies at EU Level

Within the framework of *European Union* regional policy (cf. Vanhove 1999), co-financial means from European structural funds were provided to facilitate structural change in those industrialised cities and regions eligible for help under EU regional criteria (Drerup 1997, Biehl and Hoffmann 1998). In western Germany in the period 1989 to 1999, cities and regions designated "industrial regions seriously affected by industrial decline" were given support (Schrumpf 1995). For these German *objective-2-areas*, 1.56 billion ecus[9] were made available in the programme period 1994 to 1999 (BBR 2000a, 274). A relatively wide range of measures were co-financed, in areas such as private investment, infrastructure, education and the environment (Europäische Kommission 1996). A pre-condition for the provision of structural funds is the establishment of regional development programmes. This requires a certain bottom-up-agreement regarding the selection of regional projects.

Following reunification, all regions in eastern Germany were eligible for *objective1* status, along with peripheral regions such as Portugal, Greece and Southern Italy ("regions lagging behind in development"). This status allows the broadest scope for using appropriated money and the highest award rate (max. 75 % co-financing rate). For these German objective-1-areas, 13.64 billion ECUs were made available in the programme period 1994 to 1999 (BBR 2000a, 274).

A share of the structural funds has been used for *community initiatives*. The initiatives Retex (textile regions), Rechar (coal regions) and Resider (steel regions) of the last programme period should be emphasised for their facilitating of structural change in industrialised regions. These community initiatives were set up in both western and eastern Germany.

Some changes have been made in EU regional policy for the programme period 2000 to 2006 (Europäische Kommission 2000). The new objective 2 areas ("undergoing economic and social conversion") now encompass both the old objective 2 and objective 5b areas ("development and structural adjustment of rural areas"), covering a smaller population than before. In eastern Germany, EU support under

[9] Before the introduction of the euro, the ECU served as an artificial accounting currency for calculating Community budgets.

objective 1 increased to 19.2 billion euros (BBR 2000a, 275). The aforementioned community initiatives have been discontinued.

Strategies at Federal Level

The Federal Government (Bund) supports EU regional policy by involvement in policy formulation and by co-financing measures in some programmes (mentioned below). Federal support is essential for development in eastern Germany. The spatial policies (cf. BBR 2000a, 227ff.) of the Federation and the states do not specify particular areas in industrialised cities and regions. Nevertheless these cities and regions are supported by various federal programmes.

(1) In line with the strategies mentioned above, one firstly thinks of classical *sectoral structural policy*. And indeed this policy is used in support of industrial problem areas, with a programme of subsidising hard-coal mining in western Germany being particularly important. Steel industry and shipbuilding have also been subsidised, at least temporarily, to slow the pace of structural change.

(2) Since all branches of industry in eastern Germany were equally hit by the economic crisis, sectoral structural policy as in western Germany was out of question. However the large-scale privatisation of former state enterprises necessitated the intervention of the Federation in economic development. The *privatisation agency* (Treuhandanstalt) privatised, restored to profitability or liquidated around 8,000 former GDR-enterprises with large injections of capital (cf. Kühl 1994). A heated discussion was conducted on the rescue of strategically relevant cores of industry ("industrielle Kerne", Nolte and Ziegler 1994) – for example the chemical industry in Saxony-Anhalt – and the agency doubtless contributed to the stabilisation of relevant industries. However the main task was to help create structures with the potential for long-term growth.

(3) *Regional structural policy* is next in the list of support measures for old industrialised cities and regions. The individual states of the Federal Republic have responsibility for this policy, although the Federal Government participates within the framework of the joint programme "Improvement of Regional Economic Structure" (Gemeinschaftsaufgabe "Verbesserung der regionalen Wirtschaftsstruktur"). The programme supports all regions throughout the country which are falling behind in growth and face increasing unemployment; the programme subsidises commercial and public investments, fosters business-oriented revitalisation of commercial and industrial zones and elaborates regional development concepts (Eberstein and Karl 1996/2000). The conditions for economic development are thereby improved, with old industrialised cities and regions being the main beneficiaries of the policy. The joint programme covers almost one third of the population of western Germany and Berlin. Eastern Germany is covered completely and investments are supported more intensively there (Zweiundreißigster Rahmenplan 2003, 17-18). Measures that are part of regional development concepts are given priority. The money can be combined with EU funds to finance projects.

(4) In addition to the nationwide regional structural policy, other instruments were introduced to *attract businesses* to eastern Germany. These include taxation measures (investment allowances and special depreciation), reduced-interest

loans, guarantees and risk capital funds, all directly aimed at enterprises (Prange 2000).

(5) From the viewpoint of *regional policy and regional planning*, there are two further new Federal instruments worthy of mention. The programme "InnoRegio", developed by the Ministry of Education and Research, promotes 23 innovative regional co-operation networks, mostly in eastern Germany (BMBF 2000). Other model projects of federal regional planning also applicable to eastern Germany are "Rehabilitation and Development" (Müller, Rathmann and Wirth 2002), "Towns of the Future" and "Regions of the Future" (Wiechmann 1999), serving to promote co-operation between various relevant participants. The InnoRegio and "Future" programmes started life as entries to a competition for the best development concepts. Although they are of high conceptual importance, their influence on the general development of old industrialised cities and regions is rather limited.

(6) Another spatially oriented policy concerns *urban development*, connected to soft location factors. Since the beginning of the 1970s the Federal Government and the states have provided investment aids for urban renewal and urban development in a special programme for urban policy (Städtebauförderung). In this second national joint programme, federal funds generally cover one third of the costs of urban redevelopment measures (BBR 2000c, 6). The most important objectives are the revitalisation of specific areas (mostly in or near the city centre), the preservation of cultural monuments and the development of large residential quarters with upwards of 2000 flats (cf. Walter 1997).

In addition to this classical programme for urban policy and the programme "Urban Rebuilding East" (Stadtumbau Ost), a special nationwide programme was introduced in 1999: "The Socially Integrative City" ("Die soziale Stadt") (Zweiunddreißigster Rahmenplan 2003, 13, Becker 2000). The approach here was to combine in one project the different promotion programmes and sources of funding for the elimination of social problems in rundown neighbourhoods. In general, old quarters near the city-centre or large residential areas dating back to the post-war period were designated for support. The goals in traditional industrial regions included the elimination of deficiencies in urban development, the creation of open green spaces and a general improvement in the ecological situation. Strong tax incentives for the construction of new residential buildings especially in eastern Germany have led to a new problem, gaining in importance: the vacancy of flats and residential buildings. Recently, a scheme for the demolition of buildings or upper floors has been introduced.

(7) While the regional and urban policies described above contribute to infrastructure development on a regional and urban level, federal support provides the necessary *national infrastructure*. This includes, for example, a programme for the support of universities and special public R&D facilities. The Federation also funds large-scale traffic projects in the new states, focusing on the renovation and construction of motorways, railways and waterways. Often these projects receive additional EU support. Lastly, the telecommunication sector (still partly state-owned) has invested heavily in eastern Germany.

(8) Following reunification, the headquarters of several important *national authorities* (federal offices, federal courts) as well as training and research institu-

tions were moved to eastern Germany. The most important decision in this context was surely the designation of Berlin as the capital of reunified Germany and seat of the Federal parliament.

(9) Several special regulations have been issued (Radtke and Eisenbarth 1993, Postlep 1994, Danielzyk et al. 2000) in order to meet the extraordinary challenges of *ecological remediation* in eastern Germany. 21 regionally significant large-scale remediation projects have been established, with a separate financial framework (total annual budget approx. 500 million euros). The instruments of labour market policy determined in the Employment Promotion Act (Arbeitsförderungsgesetz) showed significant improvement by 1993, with federal money granted for the labour-intensive dismantling of plants and the remediation of areas by unemployed persons. Large-scale investments for environmental improvements are supported by interest subsidies.

Together with the affected states, the federal government is also financing the immense remediation programmes for closed lignite and uranium mines; these costs would normally be borne by the mine operators. The annual remediation costs in lignite mining amount to approx. 500 million euros (1992-2002) and in 1999 created or sustained approx. 12,000 jobs (Bundesregierung 2000). Beyond the elimination of specific dangers, there also exists the possibility of purposefully adapting the newly created landscapes to some planned use (for example pedestrian greenways, vantage points for tourists). The total costs to restore uranium mining areas were estimated in 1991 at more than 650 million euros (Wismut 2000). A special programme has reduced the radiological loads imposed on humans and on the environment by uranium mining (though values have not yet returned to normal).

The basic legal, technical and financial challenges of environmental contamination and mining remediation have already been met. Urban and regional development has slackened somewhat with the relative low success of brownfield development. In spite of enormous efforts, rundown and abandoned inner-city areas exist side-by-side with commercial suburbanisation.

(10) The strategies, programmes and instruments already mentioned are intended for the development of old industrialised cities and regions. An additional important general mechanism at the federal level aims at compensating for disparities in state budgets. To ensure that "poor" states with inadequate financial resources can realise development projects (for example, the provision of state infrastructure), tax income from the Federation and the "rich" states is redistributed by a process of *fiscal equalisation* (Finanzausgleich). The fiscal transfers help such states as Saarland and Bremen which can be characterised as old industrialised. Total fiscal transfers from the Federal government to the five new states were 125 billion euros from 1991 to 1998, thus forming the second largest source of financial aid to eastern Germany (BBR 2000a, 230).

(11) The largest influx of money to eastern Germany comes from the *social security system* (pension, health and unemployment insurance). Of course cities and regions with low incomes and high unemployment rates benefit more than others. These transfers constitute a socially-oriented strategy of easing structural change.

In the framework of the unemployment insurance system, *labour market policy* is of special interest. This field of policy, which is often interconnected with structural policy on a local and regional level, aims at helping the unemployed find work in the so-called second labour market or take part in advanced training courses (for example on the basis of job creation schemes, partly in special public employment businesses). The second labour market policy has come under increasing criticism. Opponents argue that it is very expensive, and that in the long run people are better employed in the first labour market. However policy supporters say that workers in the second labour market have contributed much to the revitalisation of cities and regions by doing valuable work which otherwise would have been neglected. Eastern Germany received around 170 billion euros in the period of 1991-1998, including payments such as unemployment benefit (BBR 2000a, 230). This is by far the largest contribution of any single policy to reducing disparities between states.

Strategies at State Level

While the EU and the Federal Government concentrate on formulating and financing regional policy programmes, the 16 German states, with their close proximity to cities and regions, play a special role in implementing European and national programmes and managing funds. Moreover, they can initiate their own programmes and are responsible for the legal framework of local government policy. The states have also to ensure that fiscal transfers (kommunaler Finanzausgleich) allow municipalities with inadequate financial resources (often in industrialised regions with structural problems) to fulfil their obligations.

In the previously mentioned programme "Improvement of Regional Economic Structure", jointly run by the states and the Federal Government, the states are responsible for the programme's implementation and have the freedom to place certain emphasis on special topics. Furthermore, the states also develop their own programmes, often with the aim of integrating all available funds. The examples of North-Rhine Westphalia and the five new states are presented in the following.

The policy in *North Rhine-Westphalia* (NRW) for the Ruhr District shows how the focus in terms of policy content has shifted to the integration of sectoral measures, involving local and regional participants in the process of development and implementation of strategies (Heinze and Hilbert et al. 1996, 17ff.).

The first programme of NRW for the Ruhr District in 1968 concentrated primarily on infrastructure measures to create new jobs in industry, particularly in the field of transport. Universities and the school system were also expanded. In the programme issued two years later for the whole territory of NRW, further measures for the development of the Ruhr District were adopted. Apart from a desire to establish new firms, emphasis was placed on the improvement of the competitiveness of coal mining (introduction of new mining and processing technology). The extension of research, education and training infrastructure is considered the greatest success of both programmes. Following the Ruhr Conference, where representatives of many relevant groups were involved in discussions, the state issued the "Action Programme Ruhr" in 1979. This was a considerable step forward, including not only measures to improved qualifications, introduce innovative technologies and boost investment, but projects were also formulated in the fields of town planning, cultural policy, environmental protection and infrastructure. Since 1984, the Property Fund NRW has financed the acquisition, development and sale of selected brownfields for urban development, at the request of municipalities[10].

Facing another crisis in the coal and steel industries in 1987, the state developed the programme "Future initiative for the mining region". Although in terms of content this was a continuation of the previous programme, the approach was completely different: The idea was to ensure the greater participation of cities and regions in implementing the state structural policies. Regional conferences were used to foster a decentralised dialogue between the relevant participants. Expert knowledge and the practical ideas of local people from cities and regions could be integrated. Funds had to be used more efficiently in this co-operative approach and new projects quickly realised. The main incentive was that projects were only supported when integrated into a concept for regional development. In 1989 this regionalised structural policy was extended to all parts of NRW. Subsequently, this approach to development has proved successful in other regions in Germany.

In the same year, a second highly influential state initiative was launched: The International Building Exhibition (IBA) Emscher Park. This exhibition was designed to benefit one of the worst afflicted areas in the Ruhr District; the small river Emscher, highly polluted by industrial effluents and emblematic of the area's problems. Involving 17 cities and 2.5 million inhabitants, the exhibition ran for ten years (Faust 1999). The IBA represented a new understanding and approach to the renewal of industrial problem regions (Kilper 1999, 309). The strategic concept makes this clear: the programme's aim was to integrate, in 120 projects, various issues such as economic structure (e.g. infrastructure for innovation), ecology (e.g. landscape parks), social improvement (e.g. regeneration of urban districts), culture, history (e.g. industrial heritage, factories as the "cathedrals of the industrial epoch") and aesthetics (architecture). The projects were highly innovative and included the creation of landscape parks, the renovation of workers' housing estates, the night-time illumination of blast furnaces and innovation centres built using ecological principles.

Institutional innovations were necessary in order to make all this possible. The IBA planning corporation was specially founded to coordinate both the programme and the decentralized projects. Support for projects was allocated by way of open competition (applicants included municipalities and co-operatives). As a final point it should be mentioned that representatives from the private and public sectors have founded an "Initiative Group Ruhr" (Initiativkreis Ruhr) to promote a better image for the Ruhr District.

---

[10] The Property Fund (Grundstücksfonds) North Rine-Westphalia (established 1984) was the successor of the Property Fund Ruhr (est. 1980). Between 1980 and 1994, almost 800 million DM (mainly from the Land NRW) were used to defray the non-profit costs (UBA 1998); by 2001 more than 2,500 ha. land was bought and about 50 % developed and sold – totalling approx. 70 locations (http://www.leg-nrw.de).

Uranium Mining – Rehabilitation and Development

The "Uranium Mining Rehabilitation and Development Area" is located in the Ore Mountains of the Free State of Saxony, near the border with the Czech Republic. It embraces the towns of Johanngeorgenstadt and Schwarzenberg and five surrounding communities, with a population of 21,200 (1998). After the Second World War the Soviet-German Joint-Stock Wismut Company (SDAG Wismut) began the mining of uranium; from 1946 to 1958 a total of approx. 5,500 tonnes of uranium was extracted. In the industry's heyday in the 1950s, 115,000 people were employed at 19 pits. Structural changes in the wake of German reunification have brought high unemployment to the area, with economic development hampered by the legacy of uranium mining, i.e. mining dumps, subsidence, urban deformation and a poor image.

In order to provide assistance, the region was designated a Rehabilitation and Development Area as part of the Saxon State Development Plan, and a model project was initiated by the State of Saxony and the Federal Government. A multistage approach was adopted (Müller, Rathmann and Wirth 2002). First, a general rehabilitation and development concept was formulated. Second, the problems and potential of the region were analysed in greater detail by a steering committee, leading to the drafting of a provisional master plan. This master plan, acting as a basis for all work, comprises three elements. To begin, the project's guiding objective is set out in detail. Then seven fields of action with concrete measures are defined. These are titled – "Repairing the damage of uranium mining", "Transport", "Labour market", "Regional value-added chains", "Tourism", "Dealing with Wismut estates" and "Strengthening centre functions". Finally a concluding section addresses the issue of sustainable development, with suggestions on the enhancement of inter/intra-regional co-operation, the strengthening of "endogenous" potential, and the intensification of cross-border co-operation. Third, details regarding the implementation and funding of proposed measures are set out. The main result so far of this multistage approach was the launching of the first state-financed rehabilitation measures in 2001. Support for the action area since the project's introduction amounts to a total of 7.5 million euros. Since 1997 the basis for self-supporting development has been formed by a process of co-operation between local authorities in the framework of Regional Management, involving a steering committee and regional conferences, as well as enterprises, associations and committed individuals.

Peter Wirth, Institute of Ecological and Regional Development, Dresden

North Rhine-Westphalia, with a population of 18 million, has more inhabitants than the five new states put together. Along with this disparity in size, there are also fundamental differences in the troubles afflicting NRW as outlined above and the eastern regions. The latter are affected by state-wide de-industrialisation, with associated problems demanding urgent solutions. The new states are economically and financially much weaker, and a functioning administration system has had to be developed to handle the new situation. All these difficulties necessitated a great deal of outside help. In the first few years after reunification the eastern states received administrative help by a process of twinning with western German states.

State policy has played a relatively minor role in comparison to policy initiatives at the federal level. Nevertheless, the success of efforts to reduce disparities is very much influenced by state policy.

The following state measures can be mentioned: creation or remediation of basic infrastructures, such as traffic and education; creation of an infrastructure with direct economic benefits, such as commercial and industrial zones or investment and innovation promotion (whereby large-scale projects play an important role). Moreover, new approaches to improve the environment and landscape quality, to develop urban areas and to support cultural activities have also been adopted. The new states have also taken steps to activate regional co-operation, in particular the co-operation between municipalities (Danielzyk and Wirth 2000), in order to promote the efficient and direct use of funds by local actors. Like North Rhine-Westphalia, Saxony-Anhalt is pursuing a regionalised structural policy. Saxony has introduced a central-place system, based on co-operation between municipalities (Müller and Beyer 1999), which promote the formulation and implementation of development concepts for selected regions.

The model provided by the IBA Emscher Park has migrated eastwards. The Correspondence Region of Saxony-Anhalt for the EXPO 2000 in Hanover was organised along similar lines, based on preliminary work of the Bauhaus Dessau (Leimbrock and Lintz 2003). In 2000, the lignite-mining area of Lower Lusatia initiated a landscape-related IBA Fürst-Pückler-Land, with the aim of creating new landscapes from areas damaged by mining activities.

3.3 Regions and Cities

Strategies at Regional Level

In Germany, the regions constitute a policy level between the states and municipalities. Some planning regions are part of the spatial planning system of the states, while others, more problem oriented, are basically free co-operations between local actors (mostly municipalities); these are of growing importance for regional development (ARL 2000, BBR 1999b, Benz and Holtmann 1998). With their specialist expertise and wide range of possibilities for action, regional actors can do much to facilitate structural changes. Active competition among municipalities generally makes it necessary for superior levels (especially the states) to offer special incentives for regional or inter-municipal co-operation.

The content of regional strategies is fundamentally the same as strategies applied on a municipal level, the only difference being that they are co-ordinated and that measures are jointly implemented. Only a few examples of successful, regional co-operation in industrialised regions will be given from the great number of initiatives. For example, an institution for the joint promotion of economic development was founded in the Ruhr district for the region Emscher-Lippe, while in eastern Germany, several cities are co-operating in the development of the economic region Chemnitz-Zwickau, strengthening the field of machine and vehicle building by extending and concentrating innovation and research potentials and marketing the region as a unified site.

In many regions, actors are trying to combine the remediation of environments with the creation of areas of great beauty – beyond the necessities of economic promotion. Thus, the cities Bottrop, Essen and Gladbeck in the Ruhr area are co-operating to restore a small river system to its natural state and design a new valley landscape. The city of Leipzig and the surrounding municipalities are working together to improve the environmental situation, reclaiming derelict land to form local recreation areas and connect regional wildlife habitats. Furthermore, in an area south of Leipzig, a moonlike opencast mining district is being turned into an attractive landscape of artificial lakes, including the biggest leisure park of eastern Germany. The image of the EXPO 2000 correspondence region Saxony-Anhalt is also changing in a similar fashion. For instance, huge excavators formerly used in lignite mining have been grouped together on a peninsula to form an ensemble called Ferropolis, creating an attractive and unusual location for large open-air events.

Regional actors can choose between "hard" and "soft" institutional co-operation, depending on the degree of commitment and intensity of co-operation desired. Examples of "soft" co-operation measures are the pooling of information through meetings and regional conferences, and the joint funding of regional agencies. "Hard" co-operation implies, for example, the establishment of special purpose associations for joint performance of certain regional tasks (Zweckverbände).

Strategies at Municipal Level

Structural problems are directly felt at the level of cities. All measures must be implemented here and the effect of the measures is quickly evident. As a rule, cities mainly use traditional instruments of municipal economic development, including the establishment of new companies and support of already existing enterprises (cf. Hollbach-Grömig 1996). An important factor in economic development is the designation of commercial and industrial zones in the city's land-use plan, with appropriate land development measures and site marketing. A municipality must co-operate closely with the state if it wishes to extend the supra-local infrastructure with new long-distance railway lines, motorways and airports, or if it intends to establish large enterprises and scientific centres. A further instrument for economic promotion is the innovation-oriented creation of business and innovation centres. Also of importance are special urban programmes to foster a creative milieu and connect research, education, professional training and production to one another; this is achieved through the formation of synergistic clusters of companies and other relevant institutions.

Integrated strategies are gaining prominence in the field of urban redevelopment. Not only economic but also environmental, cultural and social aspects are taken into consideration in order to improve the soft location factors, local quality of life and the town's image. In the long term, revitalisation of brownfields and their incorporation into urban structure is seen as indispensable to further development. The same holds true for the preservation of the industrial heritage. Instead of demolishing old buildings, nowadays it is common to try to find a new use for them, and an industry-oriented tourism is even emerging. Many cities are now more highly committed to improving neighbourhoods and the natural landscape than ever before. Funds for labour market promotion are frequently used within the framework of municipal employment policy to achieve this.

Complex problems and new strategies for the promotion of economic development on the municipal level necessitate new approaches to the organisation of policy and planning. The relevant actors should be brought together efficiently and flexibly, whether administrative officials, politicians, enterprises, education and research institutions, citizens, associations of enterprises, trade unions, environmental initiatives, cultural initiatives or the media. The catchwords in this context are "discursive planning" and "co-operation", "participation" and "public-private-partnership". A good example is provided by town marketing processes, set up with representatives from all relevant parties. Privately organised associations for the promotion of economic development and location marketing are also being founded. Of course, drawing up a plan is less important than implementing concrete projects.

4 Ambiguous Perspectives

Old industrialised cities and regions in Germany, especially in eastern Germany, are undergoing a dramatic transformation, bringing challenges to all political-administrative levels. Given the tradition and constitutional aim of a balanced spatial development, these problems are being taken seriously. The affected cities and regions have been assisted with generous financial means. Strategies to tackle unemployment and environmental damage have been refined again and again, so that a sophisticated system is now in place. The many political-administrative levels involved have entailed a high (perhaps too high) degree of complexity.

At present, comprehensive approaches involving the enhancement of soft location factors such as the natural environment, culture (including industrial culture) and urban qualities are being stressed. Ultra-soft location factors such as innovative networks and creative milieus are also being examined. This implies that all relevant actors – on different political-administrative levels, in different ministries and departments, in different cities of a region and in the private sector – must coordinate their efforts.

It is difficult to assess the success of these strategies. On the one hand progress is clear and often astonishing, especially in eastern German centres such as Dresden and Leipzig. On the other hand, revitalising all the old industrialised cities and regions in Germany will be a lengthy and highly expensive project. In fact, a general convergence of living standards seems to be a remote goal.

At least two trends may change Germany's strong commitment to spatially balanced development, if only marginally. Firstly, the country's poor economic performance in the last twelve years and particularly the present stagnation have aggravated the financial problems of the public sector and diminished the scope for action in supporting structural change in backward cities and regions. The social security system faces severe pressure from demographical developments, and public debt presents a serious problem. Secondly, EU funding for regional policy in Germany may decrease considerably from 2007 on as a consequence of the EU eastward enlargement in 2004. Brussel's attention and financial means will shift to the new EU members.

Against this background, it is possible that the emphasis will no longer be on spatial equality, but rather spatial efficiency. Proponents of this shift argue that those cities and regions with the highest potential for economic growth should be the focus of support, in order to improve competitiveness on the world market and realise a higher economic growth rate for Germany as a whole.

The question is whether old industrialised cities and regions can form centres of growth. Based on the assumption that the support of eastern Germany should in principle continue, a recent study suggests that economic support should in future focus on a few agglomerations, with the decline of peripheral cities and regions being accepted as inevitable (DIW et al. 2002, 461-463). Since all agglomerations in eastern Germany are old industrialised, their assistance is assured. However old industrialised cities and regions not in the vicinity of these agglomerations would lose support.

Although it is likely that support for old industrialised cities and regions will decline overall, cities and regions still have a strong interest in their own development, motivating an even more efficient use of available support. Local and regional actors will have to learn how to exploit their individual endogenous potential, with intensified co-operation on local and regional levels to optimise development.

References

Achtundzwanzigster Rahmenplan der Gemeinschaftsaufgabe „Verbesserung der regionalen Wirtschaftsstruktur" für den Zeitraum 1999 bis 2002 (2003), Deutscher Bundestag 1999, Bundestagsdrucksache 14/776, Bonn

Albers G (ed) (1999) Planning in Germany. Introduction to System, History and Research. Outline of Ruhr Region, Frankfurt a. Main, Berlin. Special Bulletin '99 of the International Society of City and Regional Planners on the occasion of 35th Congress "The Future of Industrial Regions – Regional strategies and local action", Germering

Ambrosius G (1994) Ursachen der Deindustrialisierung Westeuropas. In: Abelshauser W (ed) Umweltgeschichte. Umweltverträgliches Wirtschaften in historischer Perspektive. Göttingen: 191-221.

ARL – Akademie für Raumforschung und Landesplanung (ed) (2000) Kooperation im Prozess des räumlichen Strukturwandels, Wissenschaftliche Planarsitzung 1999, Forschungs- und Sitzungsberichte 210, Hannover

BBR – Bundesamt für Bauwesen und Raumordnung (ed) (1999a) Aktuelle Daten zur Entwicklung der Städte, Kreise und Gemeinden. Ausgabe 1999. Berichte des BBR Bd 3, Bonn

BBR – Bundesamt für Bauwesen und Raumordnung (ed) (1999b) Perspektiven für die Region als Planungs- und Handlungsebene. Informationen zur Raumentwicklung, H. 9/10

BBR – Bundesamt für Bauwesen und Raumordnung (ed) (2000a) Raumordnungsbericht 2000. Berichte des BBR, Bd 7, Bonn

BBR – Bundesamt für Bauwesen und Raumordnung (ed) (2000b) Regionale Aspekte des wirtschaftlichen und sozialen Wandels in den neuen Ländern. Regionalbarometer neue Länder, Vierter zusammenfassender Bericht. Berichte der BBR, Bd 4, Bonn

BBR – Bundesamt für Bauwesen und Raumordnung (ed) (2000c) Urban Development and Urban Policy. An Overview, BBR, Berichte, Bd. 6, Bonn

BBR – Bundesamt für Bauwesen und Raumordnung (ed) (2001) Deutsche Einheit – die erste Dekade im Rückblick. Informationen zur Raumentwicklung, H. 2/3.2001

Becker H (2000) Das Bund-Länder-Programm "Soziale Stadt". Raumbezüge und Handlungsfelder. Die alte Stadt 2/2000: 139-149

Benz A, Holtmann E (eds) (1998) Gestaltung regionaler Politik. Empirische Befunde, Erklärungsansätze und Praxistransfer, Opladen

Biehl D, Hoffmann H-J (1998) Der Einsatz raumwirksamer Finanzmittel der Europäischen Union in den Regionen der Bundesrepublik Deutschland. Eine Bewertung aus der Sicht der Raumordnung. Informationen zur Raumentwicklung 9 (1998): 609-629

Blotevogel HH (1997) Strukturwandel und Regionalpolitik im Ruhrgebiet. In: Axt HJ (ed) Strukturwandel in Europa, Südosteuropa, Ruhrgebiet und Reform der EU-Strukturpolitik. Südosteuropa Aktuell 26. München, pp 125-140

BMBau – Bundesministerium für Raumordnung, Bauwesen und Städtebau (ed) (1991) Raumordnungsbericht 1991 der Bundesregierung. – Deutscher Bundestag 12. Wahlperiode, Bundestagsdrucksache 12/1098, Bonn

BMBF – Bundesministerium für Bildung und Forschung (ed.) (2000) InnoRegio – die Dokumentation, Bonn

Bömer H (2000) Ruhrgebietspolitik in der Krise. Kontroverse Konzepte aus Wirtschaft, Wissenschaft und Verbänden, Dortmunder Beiträge zur Raumplanung 101, Dortmund

Bönisch R, Mohs G, Ostwald W (1982) Territorialplanung. Berlin

Bundesregierung (ed) (2000): Jahresbericht 2000 der Bundesregierung zum Stand der Deutschen Einheit, Bonn

Butzin B (1993) Strukturkrise und Strukturwandel in "alten" Industrieregionen. Das Beispiel Ruhrgebiet. Geographie Heute, H. 113: 4-12

Butzin B (1996) Kreative Milieus als Elemente regionaler Entwicklungsstrategien? Eine kritische Wertung. In: Maier, J (ed): Bedeutung kreativer Milieus für die Regional- und Landesentwicklung. Bayreuth, pp 9-37

BVS – Bundesanstalt für vereinigungsbedingte Sonderaufgaben (ed) (1995) Abschlußstatistik der Treuhandanstalt per 31.12.1994

Danielzyk R (1992) Gibt es im Ruhrgebiet eine "Postfordistische Regionalpolitik"? Geographische Zeitschrift H. 2: 84-105

Danielzyk R et al. (eds) (2000) Sanierung und Entwicklung in Ostdeutschland – Regionale Strategien auf dem Prüfstand, IÖR-Schriften Bd 32

Danielzyk R, Wirth P (2000) Regionale Entwicklungsansätze in Ostdeutschland – Eine Zwischenbilanz. In: Danielzyk R et al. (ed) Sanierung und Entwicklung in Ostdeutschland – Regionale Strategien auf dem Prüfstand. IÖR-Schriften Bd 32, Dresden, pp 165-190

DIW – Deutsches Institut für Wirtschaftsforschung (ed 1999a) Gesamtwirtschaftliche und unternehmerische Anpassungsfortschritte in Ostdeutschland. Neunzehnter Bericht des Deutschen Instituts für Wirtschaftsforschung, Berlin, des Instituts für Weltwirtschaft an der Universität Kiel und des Instituts für Wirtschaftsforschung Halle. DIW-Wochenbericht 23 (1999): 419-445

DIW et al. – Deutsches Institut für Wirtschaftsforschung et al. (2002): Fortschrittsbericht wirtschaftswissenschaftlicher Institute über die wirtschaftliche Entwicklung in Ostdeutschland, edited by the Institut für Wirtschaftsforschung Halle, Halle

Drerup D (1997) German Policy Perspectives. In: Bachtler J, Turok J (Ed) The Coherence of EU Regional Policy. Contrasting Perspectives on the Structural Funds, Regional Policy and Development Series 17. London Philadelphia, pp 337-345

Dürr H, Gramke J (eds) (1993) Erneuerung des Ruhrgebiets. Regionales Erbe und Gestaltung für die Zukunft, Festschrift zum 49. Deutschen Geographentag, Paderborn

Eberstein HH, Karl H (eds) (1996/2000) Handbuch der regionalen Wirtschaftsförderung. 3. Auflage (Loseblatt-Ausgabe), Stand Juli 2000, Düsseldorf

Europäische Kommission (ed) (1996) Strukturfonds und Kohäsionsfonds 1994-1999. Verordnungstext und Erläuterungen. Luxemburg

Europäische Kommission (ed) (2000) Strukturpolitische Maßnahmen 2000-2006. Kommentare und Verordnungen, Luxemburg

Faust H (1999) Das Ruhrgebiet – Erneuerung einer europäischen Industrieregion. Impulse für den Strukturwandel durch die Internationale Bauausstellung Emscher Park. Europa Regional 7(1999): 10-18

Federal Office for Building and Regional Planning (ed) (2001) Spatial Development and spatial Planning in Germany, Bonn

Fischer A (1994) Der industrielle Strukturwandel in den neuen Bundesländern. Wirtschafts- und Sozialwissenschaften Bd. 9. Weiden, Regensburg

Fürst D, Schubert H (1998) Regionale Akteursnetzwerke. Zur Rolle von Netzwerken in regionalen Umstrukturierungsprozessen. Raumforschung und Raumordnung 56: 352-361

Georgi B (1994) Braunkohlenabbau und Landschaftshaushalt. Das Beispiel der Niederlausitz. Geographische Rundschau 46/6 : 344-350.

Grabher G (1993) Wachstums-Koalitionen und Verhinderungs-Allianzen. Entwicklungsimpulse und -blockierungen durch regionale Netzwerke. Informationen zur Raumentwicklung 3/4 (1993): 111-121

Grabher G (1994) The Weakness of Strong Ties: The Lock-in of Regional Development in the Ruhr Area. In: Grabher G (ed) The Embedded Firm. On the Socio-Economics of Industrial Networks. London New York, pp 255-278

Grabher G, Stark D (1997) Restructuring Networks in Post-Socialism: Legacies, Linkages and Localities. Oxford

Hamacher W (2001) Wohnungswirtschaftliche Alternativen versus Abrissdiskussion, Konrad Adenauer Stiftung, www.1.kas.de/publikationen/2001/ kommunal/integr_ stadtentwicklung/hamacher.html, download June 11th 2003

Hamm R (1991) Umstrukturierungsprobleme in den neuen Bundesländern und Erfahrungsmuster altindustrieller Regionen. Parallelen und Unterschiede. Raumforschung und Raumordnung, H. 2-3: 91-99

Hamm R, Wienert H (1990) Strukturelle Anpassung altindustrieller Regionen im internationalen Vergleich. Schriftenreihe des RWI Essen 48, Berlin

Härtel H-H (2001) Ostdeutschland im nationalen und europäischen Regionalvergleich. Wirtschaftsdienst H. V: 292-296

Häußermann H (ed) (1992) Ökonomie und Politik in alten Industrieregionen Europas. Probleme der Stadt- und Regionalentwicklung in Deutschland, Frankreich, Grossbritannien und Italien. Basel, Bosten, Berlin

Heinze RG, Hilbert J et al. (1996) Strukturpolitik zwischen Tradition und Innovation. Nordrhein-Westfalen im Wandel, Opladen

Henckel D (1986) Gewerbepolitik und Gewerbebrache: Aspekte der Flächennutzung im Süd-Nord-Vergleich. In: Friedrichs J, Häußermann H, Siebel W (Hrsg) Süd-Nord-Gefälle in der Bundesrepublik? Opladen, pp 51–68

Hentrich S, Komar W, Voss G, Weisheimer M (2001) Umweltschutz in Deutschland - Der Aufholprozess des Ostens. IW-Umwelt-Service Themen 2/2001, Köln

Hohlfeld P (1999) Erfolgreiche Integration Ostdeutschlands in die internationale Arbeitsteilung? Zur regionalen und sektoralen Entwicklung des Außenhandels. RWI-Mitteilungen 50 (3): 178.

Hollbach-Grömig B (1996) Kommunale Wirtschaftsförderung in den 90er Jahren, Difu-Beiträge zur Stadtforschung 21, Berlin

IWH – Institut für Wirtschaftsforschung Halle (ed) (2001) Zehn Jahre Deutsche Einheit – Bilanz und Perspektiven – Beiträge einer wissenschaftlichen Tagung, Halle

Kilper H (1999) Die Internationale Bauaustellung Emscher Park. Eine Studie zur Steuerungsproblematik komplexer Erneuerungsprozesse in einer alten Industrieregion. Opladen

Kowalke H (1998) Dresden und Lausitz. In: Kulke E (ed) Wirtschaftsgeographie Deutschlands. Gotha Stuttgart, pp 407-434

Kühl J (1994) Regionalstrukturen von Treuhandunternehmen. Stand der Privatisierung und Perspektiven. Informationen zur Raumentwicklung 4 (1994): 245-254

Leimbrock H, Lintz G (2003) "Der Ausnahmezustand" als Entwicklungsstrategie. Große umweltorientierte Kooperationsprojekte in ostdeutschen Problemregionen. Raumforschung und Raumordnung, H. 1-2/2003: 58-67

Mayr A, Taubmann W (eds) (2000) Germany Ten Years after Reunification, Beiträge zur regionalen Geographie, Bd. 52, Leipzig

MSWKS – Nordrhein-Westfälisches Ministerium für Städtebau und Wohnen, Kultur und Sport (ed) (2001) Städtebauminister Vesper gibt grünes Licht: Land kauft "Phoenix West" in Dortmund. Pressemitteilung 22. Mai 2001, Düsseldorf

Müller B, Beyer B (1999) Regionalentwicklung im kommunalen Verbund – Städteverbünde in Sachsen. Dresden, München

Müller B, Rathmann J, Wirth P (2002) Sanierung und Entwicklung umweltbelasteter Räume. Modellvorhaben in einer ehemaligen Uranbergbauregion, Frankfurt am Main

Mulke W (2000) Zehnjahresbilanz – Gespaltenes Fazit der Währungsunion. Wirtschaftsforscher sehen Fehler im Geldumtausch / Angleichung hilft Rentnern. In: Sächsische Zeitung, 28. Juni 2000: 30

Nolte D, Ziegler A (1994) Neue Wege einer regional- und sektoralorientierten Strukturpolitik in den neuen Ländern. Zur Diskussion um den „Erhalt industrieller Kerne". Informationen zur Raumentwicklung 4 (1994): 255-265

Oelke E (1998) Das Verdichtungsgebiet Halle-Leipzig. In: Kulke E (ed) Wirtschaftsgeographie Deutschlands. Gotha Stuttgart, pp 381-406

Pohl R (ed) (1995) Herausforderung Ostdeutschland. Fünf Jahre Währungs-, Wirtschafts- und Sozialunion, Berlin 1995

Pohl R (2000) Die unvollendete Transformation. Ostdeutschlands Wirtschaft zehn Jahre nach der Einführung der D-Mark. Wirtschaft im Wandel 8/2000: 222-238

Postlep R-D (1994) Organisation und Finanzierung der Altlastensanierung in den neuen Bundesländern. Altlastensanierung. Zeitschrift für Angewandte Umweltforschung, Sonderheft 5/1994: 71-79

Prange, H (2000) Lessons from financing the EU's 'silent' Eastern Enlargement of 1990, European Urban and Regional Studies 7(4): 359-369
Radtke H, Eisenbarth S (1993) Die Finanzierung der ökologischen Altlasten in den neuen Bundesländern. Umwelt- und Planungsrecht 13 (1993): 86-91.
Schaich E, Fleischhacker J (2000) The current demographic shape of Germany: Similarities and differences in eastern and western population processes. In: Kučera T, Kučerová OV, Opara OB, Schaich E (eds) New demographic faces in Europe. Changing population dynamics in countries of Central and Eastern Europe. Berlin Heidelberg New York, pp 141-162
Schmidt R (1995) Strukturveränderungen der Industrie in den neuen Ländern – Ergebnisse und Auswirkungen aus regionaler Sicht. In: Schmidt R (ed) Strukturwandel und Entwicklungsfragen Altindustrialisierter Regionen. Ergebnisse, Probleme und Chancen eines ökologisch verträglichen Strukturwandels von Industrieregionen in Sachsen, Sachsen-Anhalt und Thüringen. IÖR-Schriften Bd. 13, Dresden, pp 6-15
Scherf K, Scholz D et al. (eds) (1990) DDR – Ökonomische und soziale Geographie, Gotha
Scholz D (1994) Wirtschaftsräumliche Strukturveränderungen in den neuen Bundesländern Sachsen und Thüringen zu Beginn der Neunziger Jahre. Sitzungsberichte der Sächsischen Akademie der Wissenschaften zu Leipzig. Philolog.-hist. Klasse Bd. 133, Leipzig
Schrader M (1998) Ruhrgebiet. In: Kulke E et al. (Hrsg) Wirtschaftsgeographie Deutschlands. Gotha Stuttgart, pp 435-463
Schröter HG (2000) Von der Teilung zur Wiedervereinigung (1945-2000). In: North M (ed) Deutsche Wirtschaftsgeschichte. Ein Jahrtausend im Überblick. München, pp 351-420.
Schrumpf H (1995) Die Förderung altindustrialisierter Gebiete durch die Europäische Union (Ziel-2-Förderung) und die Gemeinschaftsaufgabe "Verbesserung der Regionalen Wirtschaftsstruktur". In: Ridinger R, Steinröx M (eds) Regionale Wirtschaftsförderung in der Praxis. Köln, pp 98-118
Sinn, H-W (2002) Germany's Economic Unification: An Assessment after Ten Years. Review of International Economics 10(1): 113-128
SRU – Rat von Sachverständigen für Umweltfragen (1995) Sondergutachten II "Altlasten" des Rates von Sachverständigen für Umweltfragen. Bundestagsdrucksache 11/6191, Bonn
Statistisches Bundesamt (1999) Tabellensammlung zur wirtschaftlichen und sozialen Lage in den neuen Bundesländern. Arbeitsunterlage 2/99. Wiesbaden
Statistisches Bundesamt (2000) Statistisches Jahrbuch 2000 für die Bundesrepublik Deutschland. Wiesbaden
Steiner M (1985) Old Industrial Areas: A Theoretical Approach. Urban Studies H. 22: 387-398
SVR – Sachverständigenrat zur Begutachtung der gesamtwirtschaftlichen Entwicklung (1999): Wirtschaftspolitik unter Reformdruck, Jahresgutachten 1999/2000, Wiesbaden
SVR – Sachverständigenrat zur Begutachtung der gesamtwirtschaftlichen Entwicklung (2002): Jahresgutachten 2002/2003, Bundestagsdrucksache 15/100 vom 15.11.2002
Tettinger PJ, Mann T (1994) Industrieflächenrecycling und Altlastensanierung. Altlastensanierung Sonderheft 5 (1994): 63-70
Tschentscher A (2002) The Basic Law (Grundgesetz). The Constitution of the Federal Republic of Germany, Würzburg
Turowski, G (ed) (2002) Spatial Planning in Germany. Structures and Concepts, ARL, Studies in Spatial Development, Nr. 1, Hannover
UBA – Umweltbundesamt (ed) (1998) Revitalisierung von Altstandorten versus Inanspruchnahme von Naturflächen. UBA-Texte 15 (1998)

UBA – Umweltbundesamt (ed) (2000) Handlungsempfehlungen für ein effektives Flächenrecycling. UBA-Texte 10 (2000)

Vanhove N (1999) Regional Policy: A Europaen Approach. Aldershot Brookfield USA Singapore Sydney

Walter K (1997) Entstehung und Implementation der Städtebauförderung im bundesstaatlichen System: Eine empirische Untersuchung des föderativen Konflikt- und Konsensverbundes. Berliner Schriften zur Demokratieforschung 5, Frankfurt/M

Wienert H (2000) Probleme des sektoralen und regionalen Wandels am Beispiel des Ruhrgebiets, Beiträge der Fachhochschule Pforzheim Nr. 91, Pforzheim

Wismut GmbH (2000) Finanzierung. In: http://www.wismut.de/ (12.12.2000)

Wood G (1997) Social and Economic Restructuring in Old Industrial Areas: A Comparison of North-East England and the Ruhr District in Germany. In: Blotevogel HH, Fielding AJ (ed) People, Jobs and Mobility in the New Europe. Chichester. pp 189-215

ZK der SED – Zentralkomitee der Sozialistischen Einheitspartei Deutschlands (1989) Analyse der ökonomischen Lage der DDR mit Schlußfolgerungen – Im Auftrag des Generalsekretärs des ZK der SED Egon Krenz, 27.10.1989. GVS b 5 - 1155/89, Manuskript

Zweiunddreißigster Rahmenplan der Gemeinschaftsaufgabe „Verbesserung der regionalen Wirtschaftsstruktur" für den Zeitraum 2003 bis 2006, Bundestags-Drucksache 15/861 vom 10.04.2003, Bonn

Hungary: An Umbrella Plan for the Regions

Elisabeth Visy, György Zala and Krisztian Schneller

VATI Institute of Regional Planning, Budapest

1 Historical Development

The so-called "reconciliation" between Austria and Hungary in 1867 provided the impetus for an industrial upswing in both countries. Political consolidation and the strengthening of the Hungarian economy stimulated industrial growth. In this first promising period of modern industrialisation the spatial pattern of heavy industries (mining, energy production, iron and steel industries) was determined by the location pattern of energy and mineral resources. Outside the capital Budapest (formed in 1872 by linking the cities Buda, Pest and Óbuda) most of the industrial centres experiencing rapid industrial growth with associated improvements in education were developed in the Transylvanian and Highland regions (Cluj, Satu Mare, Kosice, Bratislava, Banska Bistrica etc.). These regions and their centres were seized from Hungary after World War I.

During the difficult process of consolidation in the inter-war period a new industrial pattern slowly evolved. Industrial development centred in Budapest and surrounding industrial towns (in 1950 incorporated into the capital to form Greater Budapest). This pattern was later completed by the development of a few secondary centres. As the country had been forced to give up the most valuable areas of mineral resources, the inter-war process of industrialisation was based on those resources still available (primarily bauxite in the Middle-Transdanubian Mountain Range), on basic agricultural products (textile and food industry) and on industries requiring an input of skills and knowledge rather than raw materials. The latter trend resulted in the development of electric, electronic and other specialised industries in Budapest and in some medium-sized and small towns (for example china and pottery in Pécs, Herend and Hollóháza; printing in Gyoma; lace in Győr; paprika mills in Szeged and Kalocsa; coach building at Kócs; yacht building in Balatonfüred etc.; crystal in Ajka; leather in Tata; paper mills in Balatonfűzfő and Diósgyőr). In the late 1930s war preparations stimulated the development of heavy industries and the search for exploitable energy resources.

A new spurt of accelerated development in heavy industry began in the late 1940s and early 1950s as a result of Soviet models being applied to all newly socialised countries. An inherent contradiction in this development was the emphasis on heavy industry in the national economic structure, in disregard of the domestic economic potential. Apart from wartime measures, there was no long tradition of heavy industrial development in Hungary (viewed within its current borders). There were no mineral resources, either for raw material (iron and steel) or for energy. Therefore all energy and mineral resources had to be imported, primarily from the Soviet Union, with most products then being exported back. The concentration on the growth of iron and steel production and heavy manufacturing was highly detrimental to the survival of previously promising and more sophisticated sectors, such as electronics and precision instruments. These and other industries (textile, glass and pottery, etc.) saw a decline, and all that was left was directly subordinated to heavy industries. For instance, the output of the traditional china and pottery plant in Pécs was shifted to the manufacture of industrial china for use in electricity distribution. The dominance of heavy industries and mass production in all economic sectors led to an overall decline in the quality of products and services.

In Hungary regional disparities have been a major development issue throughout the 20th century. A fundamental element of spatial disparity has been the gap between the capital, the capital region and the rest of Hungary. During the four decades up to the end of the 1980s changes in regional structure took the following pattern:

- the density of people, activities and investments in Budapest increased;
- centres of heavy industry (mining, metallurgy and heavy manufacturing) evolved on the basis of existing towns and new towns, together forming a northeast-southwest axis of industrial development;
- from 1960 a tendency towards decentralisation was seen, increasing further in the 1970s to involve all county centres and major towns in the development of industry and housing;
- a modest decentralised development of services took place in the 1970s, especially of secondary and higher education, resulting in the establishment of institutes of higher education in all county centres;
- industrial decentralisation spread further in the 1970s, with the establishment of minor industrial assembling plants in small towns and some rural settlements;
- industrial decentralisation also took the form of decentralisation of labour intensive industries, bringing about the establishment of auxiliary firms[1] without local management;
- all the decentralisation efforts resulted in low-quality developments in housing and industry, while the concentration of capital and intellectual resources was maintained in the capital and a small number of regional centres.

In the 1960s an experimental reform was drafted, leading to the so-called "new economic machinery" (introduced in 1968). It was intended to decrease the role of

[1] In the 1970s and 1980s – as part of the deregulation process – the farming cooperatives were allowed to establish firms within their own organisation whose activity was unrelated to agricultural production. Their activities ranged from processing and marketing agricultural products to cleaning offices and manufacturing electrical equipment.

central government directives, with the national economic plan determining merely the main directions of economic development. Firms were not given detailed instructions and had a broader scope of action in autonomous decision-making. Incentives were introduced to ensure that the firms' production figures were in line with guidelines set down in the national economic plan. The actual implementation of this policy did not fulfil the original intentions. The characteristics of indirect control gradually weakened in the following years. Nevertheless, some scope for entrepreneurial activity persisted during consecutive periods of decline and upheaval. From the early 1980s economists, sociologists and middle-level managers were increasingly convinced of the necessity of consistent economic reforms, finally leading to the recognition of the unavoidability of fundamental political change.

2 Present Situation of Industrialised Cities and Regions

Structural change was shaped and determined by global and internal trends, with painful recovery measures being necessary to transform an essentially weak and flawed national economy. One can say that the sharp decline in the national domestic product during the early and middle 1990s was the consequence of a surgical procedure carried out to create favourable conditions for future recovery. Annual GDP fell sharply after 1989, reaching a low in 1993 of only 82 % of the 1989 figure. GDP increased slowly thereafter, returning to the level of 1989 only in 2000.

The impact of industrial restructuring and the decline of heavy industries and mass production were first felt in the 1980s. Redundancies initially affected the unskilled workforce living in rural peripheries and commuting to industrial centres to work. The second wave of industrial decline saw the closure of small auxiliary firms in the rural centres. Although these processes in the 1970s and 1980s prepared some of the ground for change, the subsequent political, economic and social changes in the 1990s were much wider and more radical. Important elements such as privatisation, the loss of the Soviet and eastern block markets, new trade orientations and cross-border relations, and the arrival of foreign capital, have all contributed to a rearrangement of the spatial structure and have changed the pattern of regional disparities. The nature and degree of spatial changes means that the polarisation of regions has become more visible than ever before.

The eastern and north-eastern regions have been, on the whole, the losers of the transformation of the political system, suffering long-term unemployment and a low rate of foreign investment. Although the number of enterprises is high, poor transport connections to the dynamic regions and the capital city ensure that they have difficulties in maintaining healthy profits. Only a few larger cities and county centres are economically successful, and these have a tendency to absorb resources from surrounding areas. They attract both capital and highly-skilled workers, and show a high concentration of government and non-government institutions.

Map: Hungary – Former industrial regions

The overall economic decline had obvious causes, common to all east European countries. The loss of the eastern block markets meant a loss of exports in heavy industry, also affecting other industries and agriculture. The economic consolidation after this drastic change took some time, but the fact that the Hungarian national economy was more open to western economies helped to decrease the time needed.

The influx of foreign capital was sometimes detrimental to economic recovery. The abbroachement of Hungarian firms (sugar, textile etc.) often meant the abbroachement of the market rather than production: several firms under the control of foreign owners were closed down, contributing to the growth of unemployment and a decline in national economic output. The overwhelming bulk of foreign capital has been attracted to central Hungary (Budapest, Pest County) and in some counties in the north-west (Győr-Moson-Sopron, Vas, Fejér). Government subsidies and tax incentives are provided to increase the attractiveness of some regions. This is the reason for the slightly improved position of the north-eastern counties (Borsod-Abaúj-Zemplén and Hajdú-Bihar). The instability of the neighbouring countries south of Hungary has prevented foreign capital investment in the southern regions.

Spatial differences in industrial production have evolved by the end of the 1990s (see Table 1). The per capita values are by far the highest in the counties of western Hungary (e.g. Fejér, Győr-Moson-Sopron and Vas) where foreign capital was attracted first and where the growth rate of multi-national and national firms was rather high. The per capita figures in other dynamically developing areas such as Budapest and its metropolitan region (Pest County) and Veszprém County are lower. In these regions primary growth is in services. The low per capita values in the former industrial regions (Borsod, Nógrád, Heves, Baranya, see map) are clearly due to industrial decline.

In the dynamic industrial regions growth has continued in the last years of the 1990s. Some of the previously declining counties are slowly catching up, for example Baranya, Zala, Somogy and Tolna in South Transdanubia. In the counties of Bács-Kiskun and Nógrád growth only began a couple of years ago and acceleration is due to their vicinity to the Budapest Metropolitan Region, helped by improvements in accessibility (construction of the new motorway sections M5 and M3). The rate of growth was also considerable in the county Szabolcs-Szatmár-Bereg, the poorest and most underdeveloped county in north-eastern Hungary. The county was the first recipient of both national and international support. Industrial decline is continuing in the counties of Borsod (previously the main area of heavy industry), Heves and Békés (underdeveloped rural county). Trade and diverse services are growing faster than industry in Hajdu-Bihar and Csongrád.

While the unemployment rate saw a gradual decrease, especially since 1995, to the level of 1998 shown in Table 1, the reasons of this decrease are varied. In the dynamically growing regions economic improvement led to a better employment situation. However in less developed eastern counties unemployment still remained high. Little improvement was seen in the former heavy industrial counties such as Borsod and Nógrád. The relatively low unemployment figures in some counties are due to an ageing population rather than economic recovery.

Table 1. Industrial Production, 1998

Capital, County	Industrial output per capita, Average=100 %	Unemployment rate, %, Dec. 1998
Budapest	86	3.9
Bács-Kiskun	55	9.4
Baranya	54	11.0
Békés	58	12.3
Borsod-Abaúj-Zemplén	79	17.7
Csongrád	9	7.8
Fejér	340	7.6
Győr-Moson-Sopron	238	4.4
Hajdú-Bihar	59	13.7
Heves	77	11.6
Jász-Nagykun-Szolnok	81	12.4
Komárom-Esztergom	131	9.5
Nógrád	55	14.7
Pest	74	5.9
Somogy	52	11.2
Szabolcs-Szatmár-Bereg	47	16.9
Tolna	72	11.5
Vas	195	5.2
Veszprém	79	7.9
Zala	99	8.0

Source: Regional Statistical Yearbook (1999)

In 1999 industry was the leading economic sector in the dynamic western Hungarian counties like Győr-Moson-Sopron, Vas, Fejér, Komárom-Esztergom etc. In the central region (Budapest and Pest County) services contribute the highest share to the employment structure. One positive development is the changing behaviour shown by multinational enterprises. Initially foreign firms were primarily involved in low-tech and low-skilled production (e.g. product assembly). Some of these firms have recently relocated to countries providing cheaper labour and have been partly replaced by more sophisticated and stable firms. Whereas firms used to bring in their own management and experts, now more and more Hungarian specialists and managers are given top jobs or are actively involved in management. One consequence of high incomes and activity rates in these dynamic counties is that people are generally less motivated to continue in higher education and re-training programmes. In lagging regions, where prospects of productive employment are less secure, young and middle-aged people are more willing to turn to higher education and further training to improve their chances of employment.

The county-based analysis can be complemented by an analysis of selected industrial centres. Table 2 shows their population development. It should first be pointed out that the overall population in Hungary decreased in the last decade, due to a low birth rate and relatively high death rate amongst the middle-aged. Furthermore, population in cities has decreased because of a migration of families (especially young families) into outlying suburban settlements. Therefore the rate of population decrease viewed in isolation is not necessarily a good indicator of a city's development. However, in agreement with the analysis above, it can be said that the cities located in the former industrial regions show – with the exception of Budapest – the biggest decline in population (e.g. Miscolc, Salgótarján, and Pécs).

Table 2. Population and population change in selected industrial centres

City, town	Population 1998 (1,000s)	Percentage of 1990 figure %
Budapest	1,839	91.2
Ajka	33	96.1
Bátonyterenye	15	97.4
Dorog	13	101.5
Dunaújváros	56	95.0
Győr	127	98.4
Kazincbarcika	34	92.0
Komló	28	94.6
Miskolc	174	88.4
Oroszlány	21	99.5
Ózd	41	93.3
Pécs	159	93.3
Salgótarján	44	92.9
Székesfehérvár	105	96.6
Tatabánya	72	97.0
Tiszaújváros	18	95.6
Veszprém	63	98.1

Source: Statistical Yearbook of Hungary (1998)

In detail, the following groups of selected industrial centres can be distinguished. Budapest is the capital city and the only centre with a population over one million. It is also the only city in Hungary to enjoy regular international air transport. Job losses from industrial decline have been offset by the rapid growth of services, and derelict industrial sites have been renovated for use by commercial facilities. Nevertheless there are still signs of dereliction, with employment shortage in certain sectors, and growing numbers of unemployed and homeless people coming from less developed parts of the country and from abroad in search of work.

Győr and Székesfehérvár are dynamic centres of growth. Their further development largely depends on their potential to maintain and promote the existing high quality industries, while attracting more service-based industries with an attractive environment and high quality workforce.

Wise local management and a wish to ensure locational advantages led to the consolidation of two centres of heavy industry, Tatabánya and Dorog (Komárom-Esztergom county). In both cases a viable future was assured by the attraction of small and medium-sized firms to balance industrial closures.

The towns of Dunaújváros, Tiszaújváros (Borsod-Abaúj-Zemplén county) and Kazincbarcika (Borsod-Abaúj-Zemplén county) were established as socialist new towns in the 1950s and 1960s. Their economic structure has survived, at a reduced level, by the continuous modernisation of iron and steel production at Dunaújváros and of chemical industries in the other towns, further supported by a process of industrial diversification in the period preceding 1990. However their long-term prospects are uncertain.

Although heavy industry used to dominate the economic structure of Pécs and Veszprém (coal and uranium mining in Pécs, chemical and machine industry in Veszprém), commercial, administrative and cultural services also had an important, if subsidiary, role. These service industries are now the prime focus of development efforts. Both cities could profit from their cultural and architectural assets as well as their natural potential for tourism.

Komló (Baranya county), Bátonyterenye (Nógrád county) and Oroszlány (Komárom-Esztergom county) were small but thriving coal-mining towns which, with the closure of the mines, have entirely lost their means for existence. Their population is ageing, and most live on subsistence support. Migration from these towns is restricted by the demographic structure, by low housing prices and the scarcity of alternative employment in the neighbourhood.

The situation is similar in Ajka (Veszprém county) and Ózd. Both were historic centres of metallurgy (aluminium production at Ajka, iron and steel production at Ózd). As they were not socialist new towns they did not figure as priorities of government support when the first difficulties of heavy industries emerged in the late 1970s, and little if any effort was made to improve their situation. From 1990 the economic crisis hit these areas with particular severity. Subsequently a number of small and larger scale experiments were attempted to save these areas, including the continuation of production under local management, the support of new firms, and a relocation of some of the workforce to dynamically developing regions. After disappointing results new solutions are emerging through public-private partnerships.

The last centres to be described are Miskolc and Salgótarján. These county centres were major centres of iron and steel production, making them employment centres of outstanding importance and ensuring large-scale support for urban development. Their industries provided a relatively high salary and sufficient housing for the local workforce, so that both cities were centres of in-migration. However, as the general quality of life was often below average, continual out-migration was also substantial. Thus growth was associated with a substantial fluctuation of the population. Once the attraction of industrial jobs began to wane the loss of population began to dominate. Nevertheless, in view of the size of their population and considerable industrial workforce, the government was reluctant to face the problems of sudden, large-scale unemployment. The operation of industrial firms continued with government support, delaying the loss of jobs and slowing down the processes of change. By the end of the last decade this arbitrary maintenance of industrial production was no longer feasible. Meanwhile, alternative strategies, though worked out under various international programmes (Phare and bilateral programs), were not implemented.

Both cities have features which could support the restructuring process. They are access points to areas of outstanding natural beauty and have sufficient historic monuments and natural resources (spas etc.) to attract tourism. Both cities are close to the national border with Slovakia, giving them access to funds in support of cross-border cooperation. There are some local enterprises as well as civil organisations and there is a tendency to self-help and a commitment to change. The solutions in both cities – in view of the size and character of restructuring processes – must be sought in diverse ways.

The attraction of domestic and foreign capital is promoted by the national government. New industrial parks have been established and are now gradually attracting business. A range of small and medium-sized firms has emerged, more or less successfully, though often held back by poor marketing. Cultural institutions (museums, theatres etc.) are seen as important motors of city promotion. Educational institutions (schools and the university in Miskolc; schools, polytechnics and training centres in Salgótarján) are essential focal points. After the failure of industrial cooperation between businesses in Miskolc and Kosice the first steps in cross-border cooperation are now being undertaken in the fields of environmental protection and tourism, with joint programs being organised.

Consolidation in Miskolc and Salgótarján will be a long and painful process, and it can only be hoped that the current phase is the beginning of a period of improvement.

3 Strategies

3.1 National Spatial Development Strategy

A National Spatial Development Strategy was drawn up and approved by the Hungarian Parliament in 1998 to help promote co-ordinated growth and development as well as providing a framework for development programmes which implement international, national and local support measures.[2] This strategy addresses the main issues of regional and urban development.

The *vision of spatial development* is that of a spatial structure which allows Hungary to act as a bridge between countries in the community of European states lying on the east-west axis and the north-south axis. This role is in accordance with Hungary's cultural heritage, geographical features and economic development. Regions with different socioeconomic backgrounds should proceed along their own distinctive and specific course of development, with a thorough division of labour based on mutual accord rather than subordination. Territorial disparities and the number of disadvantaged regions falling behind in socioeconomic terms should be reduced. Budapest is seen as a leading and dynamically developing centre within central Europe, co-operating intensively with surrounding regions and showing a well-balanced division of labour with other parts of the country. The regions of the country, together with the border regions of neighbouring countries, will be the key co-ordinators of European and cross-border co-operation, forming an arena of close co-operation in the area of mutual development policy. Hungary's urban axes and transport corridors should be closely integrated into the

[2] The following description of the National Spatial Development Strategy is mostly based on direct quotations.

European spatial structure, thereby creating dynamic elements of spatial development. The development of innovation and entrepreneurial centres, as well as industrial parks and business areas, will contribute to a more balanced spatial structure. The urban network must be well balanced, and each settlement should have access to the public services of urban institutions. A decentralised system of intellectual and educational/training centres would provide the driving force for regional development. Co-operation on the level of the sub/micro-regions (NUTS 4)[3] is supposed to be the primary level of the settlement network, with multiple division of labour between individual settlements. Special recreational areas are thought to be the main attractions for international and domestic high-quality tourism, and here special emphasis is placed on the conservation of landscape and environment. Established nature conservation areas preserve natural and landscape features which reflect the identity of the region, demonstrating the responsibility of today's generation for all following generations. The condition of the environment should be of primary importance, with the conservation and good management of natural resources enjoying strategic priority.

Parliament established the legal framework for a decentralised system of institutions by passing the 1996 law on spatial development and planning (Regional Development and Planning Act, XXI/1996),thereby declaring that those responsible for spatial development must co-operate on the basis of full partnership.

The *general aims of spatial development* refer to the country as a whole, as well as treating specific spatial disparities. The general aims in regard to the first aspect are the following:

- promotion of the cultural and financial welfare of citizens and the nation; the development to a social market economy; stable economic growth; improved competitiveness and the establishment of conditions for sustainable development;
- promotion of the spatial spread of innovation, increasing the fund-raising capability of settlements, counties, regions and strengthening their ability to contribute to development projects; development of a spatial structure conforming to social, economic and environmental goals; the promotion of a coherent spatial structure and settlement network over the country;

General developmental aims regarding the alleviation of territorial disparities are:

- reduction of territorial disparities between regions, counties, the capital, the countryside, cities, towns and villages, developed and underdeveloped regions and settlements, as well as between eastern and western Hungary, in their conditions of life, i.e. economic, educational, cultural, health care, social and infrastructure conditions;
- prevention of new crisis areas emerging and equal social opportunities.

[3] According to Eurostat there are 7 NUTS 2 regions (Tervezesi – Statisztikai Regio), 20 NUTS 3 regions (Megyek/counties and Budapest) and 150 NUTS 4 regions (Statisztikai Kisterseg) in Hungary.

The *overall national guidelines of spatial development* are intended to ensure the harmonisation of social solidarity and long-term economic interests, the implementation of the principles of sustainable development, the spatial expansion of innovation, the development of a balanced spatial structure, and the sustainable use of resources.

There are special long-term *development aims and priorities for regions* in Hungary. These regions are divided into disadvantaged regions, regions of specific natural and geographical features, and border regions. The aims and priorities in these regions vary according to their different potentials.

The *disadvantaged regions* comprise four types. In regions with a socioeconomic lag, internal investments and foreign capital investments need to be encouraged to create jobs. Those regions have to establish the necessary infrastructure to promote economic development. Furthermore, human resources must be enhanced, basic social and health services improved, and support given to complex projects which develop recreational tourist features and attractions. Regions undergoing industrial restructuring are most interesting here. Foreign capital investments must be encouraged, with the promotion of industrial restructuring and innovation, the organisation of training and retraining programs, and improvements made in infrastructure, with special regard to environmental investments. Regions of agricultural rural development will be assisted by farm subsidies in less favoured agricultural regions, the integration of agricultural production (processing, marketing and related infrastructure developments), and support given to supplementary part-time jobs for agricultural workers and other employment opportunities. The local economies of regions facing severe long-term unemployment need development, with the introduction of part-time employment and public work programmes, as well as training/retraining programmes, improved social services and mental health programmes.

The aims and priorities for *regions of specific natural and geographical features* comprise:

- landscape conservation and the use of land in harmony with the natural carrying capacity, as well as a balance and coherence between the natural and man-made environment and the containment of urban sprawl;
- the development of environmentally friendly tourism by creating the necessary infrastructure for eco-tourism in conservation areas, with a harmonisation of agricultural production and nature conservation;
- the reduction of damaging processes in environmentally endangered regions and the prevention of further environmental damage by appropriate waste disposal and sewage treatment; reduced air pollution by the construction of bypass roads for urban and protected areas, and implementation of a national environmental rehabilitation programme.

The following aims and priorities were formulated for *border regions*: co-ordination of common spatial development programmes, co-operation in joint ventures such as industrial parks and business areas, improvement of the quality and quantity of border-crossing points, implementation of joint environmental projects and the common development of an information infrastructure.

General economic policy must be brought in line with the spatial development policy aims. Tasks set out in the National Spatial Development Strategy must be taken into consideration when preparing annual budgets, in sectoral and spatial development programmes, and in new legislation. The earmarking of various government funds for certain projects must be coordinated, and other central financial instruments and benefits must also be adapted to national strategies. Budgetary reform must support the desired territorial equilibrium by regulating local authorities and, if justified, the financial instruments and regional resources available to decentralised institutions should be increased. A system for co-ordinating the distribution of spatial development resources must be developed. Auditing and financial control systems need to be further developed in order to better assess project proposals, while making the use of spatial development resources more transparent and accountable. Programme financing should be introduced gradually, with priority given to tenders for development projects which are related to spatial development. The Hungarian Development Bank and the network of regional development associations funded by the Investment Bank and by local authorities should assume more vital roles in the promotion of the regional economy. Financial services contribute greatly to the development of regions and these should be fostered in line with the development of the banking system and through the joint efforts of the state, local authorities, chambers of industry and the Hungarian Foundation for the Development of Enterprises.

In order to develop the *regional institutional system*, special regions must be designated whose co-ordinated planning and development are of national interest. The co-operation of county and regional development councils should be encouraged, and the planning and co-ordinating activities of the latter should be extended by involving city representatives and associations of NUTS 4 units, thereby cutting across the borders of neighbouring planning-statistical regions and territorial chambers. A monitoring system should be developed for the planning-statistical regions on the basis of information derived from county level.

The National Development Strategy and the regional programmes deal with the issue of industrial restructuring and its impact on former industrial centres, suggesting various approaches for the implementation of proposed operational programs. The current considered optimism is explained by the slow but visible process of dynamic development spreading towards eastern and southern Hungary. This process is supported by improved relationships with neighbouring countries.

The Successful Transformation of Tatabánya

Tatabánya is a former mining town with a population of 75,000. It is situated in the Transdanubian mountains, midway between the two most important development centres in Hungary in the 1990s, namely Budapest and Győr. The M1 motorway and the main railway line (Vienna-Budapest) pass through here, providing excellent transport and communication connections. Coal-mining and related energy sectors dominated the economic history of the 40 year-old socialist town. After 1990 the mines and connected industries were closed. Investment in the city and its surrounding region fell dramatically and the unemployment rate jumped to almost twenty per cent, forcing many workers to leave the city. The town administration soon realized that they could not count on direct help from the central government.

The business and civic leaders of Tatabánya worked with the mayor to establish an Economic Development Organisation; this soon drew up a long-term economic development strategy to deal with the depression. A Canadian consultant company gave important help in the form of a city management programme. The principal objective of the plan was structural change, specifically a transformation of industry. The aim set forth in the strategic plan was the liquidation of the old economic structure and the formation of a new structure capable of renewal. This meant that the coal-based heavy industry had to give way to modern electronics, new technologies and environmentally-friendly industry.

The education level of local workers was insufficient, so training and further education plays an important role in the strategy. In 1992 several local actors established a privately owned institute for higher education (College of Modern Business Sciences). The proximity of Tatabánya to Budapest, the good transportation connections, availability of labour and the existing industrial tradition provide a suitable environment for new business. An "industrial park", with basic infrastructure (communication and transport facilities, power and water supply etc.), has been established for new investors. This development won an Industrial Park Award, helping to tap other governmental sources of funding for further projects. By 1999 more than 20 companies had invested around EUR 300M in Tatabánya, creating more than 6,000 jobs and having a significant impact on socio-economic conditions in the town.

The Economic Development Organisation and the Municipality of Tatabánya play active roles in the development process. Apart from the aforementioned measures, tax exemptions and tax allowances are used to support new enterprises, provide investment and help to create jobs, as well as being used to finance necessary marketing activities such as the production of promotion material and participation at exhibitions and fairs. Economic growth and a gradual easing of the crisis situation seen in the past few years can mainly be attributed to quick and flexible management decisions. The most hopeful approach is that the town can use its favourable economic geographic features in the North Transdanubian area to become a regional sub-centre, with connections to the developing cluster of motorcar manufacturers and electronic industries in the area.

András Grosz, West Hungarian Research Institute, Győr/Hungary

3.2 Széchenyi Plan

The Széchenyi Plan (Hungarian Government 2001), named after István Széchenyi, a prominent activist for social and economic reforms in the 19th century, is a national medium-term economic development programme for the period 2001-2006 which allocates funds for activities in priority fields of policy.[4] The plan was prepared by the Ministry of Economic Affairs as an umbrella of specific economic development programmes, with the focus being on direct and indirect conditions of economic development. Individual programmes are mutually supporting. The plan provides a framework for sectoral, functional and regional programmes and aims to:

- ensure rapid and sustainable economic growth;
- widen and strengthen the bases for growth;
- develop economic growth centres;
- support integration into Europe;
- mobilise domestic and foreign economic resources.

In order to ensure rapid and sustainable economic growth the Széchenyi Plan must ensure that the bases of economic growth have sufficient width and depth for the various sectors, industries and regions. The goal is to involve new sectors, business circles and regions in economic growth. For this reason the plan focuses on the development of those sectors such as tourism and R&D where Hungary has clear natural advantages and a promising potential for development. SMEs are given support, with production, innovation and information links between modern large export-oriented companies and domestic SMEs strengthened in order to extend growth into different sectors. Development of the transport infrastructure – expressways, railways and regional airports – means that FDI can be channelled into former industrial and less developed regions. With the implementation of independent regional development subprogrammes, the Széchenyi Plan endeavours to foster growth across the whole country.

The various development programmes and subprogrammes included in the Széchenyi Plan are as follows:

- *Enterprise Development Programme*, with subprogrammes for SME development, bridging and investment promotion.
- *Housing Programme*, with subprogrammes to increase and modernise the housing stock, help people obtain dwellings and increase mobility by modernising the housing loan system, as well as aiming to enlarge the rented apartment sector and mitigate certain social problems.
- *Tourism Development Programme*, with subprogrammes for the development of health care and spa tourism, convention tourism, theme parks, heritage tour-

[4] The following description of the Széchenyi Plan is mostly based on direct quotation. In 2003 the Plan was superseded by the Europa Plan, which differs mainly by placing even greater emphasis on the accession process.

ism, tourism-related information technology systems, and various tourism-related quality products.
- *Programme for the Support of Research, Development and Innovation*, with subprogrammes for a National Research and Development project, extension of existing R&D support schemes, and increase in the absorption capacity of the R&D institutional network.
- *Information Society and Information Economy Development Programme*, with subprogrammes for government development, greater availability and access of up-to-date information and communication technology, the foundation of an electronic economy, emphasis on the culture and content development of information, and quality of life and awareness-raising.
- *Expressway Development Programme* and related subprogrammes for infrastructure developments, such as the M3 expressway between Füzesabony and Polgár, reconstruction and extension of the M7 expressway, the Szekszárd Duna bridge and connecting freeways, modernisation of rail tracks, regional airport development, flood prevention and groundwater control.
- *Regional Economy Development Programme*, with subprograms for the development of regional innovation systems, the establishment of regional clusters, the development of regional electronic markets, the creation of model programmes for economic development in micro-regions, and the development of uniquely Hungarian products with good market access.

The Széchenyi Plan is firmly based on a market economy approach and is geared towards EU practices in structural policy. Funding is allocated to programmes and projects with highly specific and closely examined objectives. The word "plan" in the title, though echoing forty years of state socialist planning is, of course, something completely different. It is important to ensure that the financing of the programme is such as to prevent an easy distribution and consequent frittering of resources and funds. The Széchenyi Plan is constructed around an extremely strict system of objectives and means. This complex economic development plan can be taken as a starting point for the use of state resources; the challenges facing the Hungarian economy are clearly stated, as are the requisite development programmes and subprogrammes to meet such challenges. The plan elaborates individual measures according to the requirements of successful programme implementation, assigning resources to these measures. Each programme included in the Széchenyi Plan contains both a general overview and detailed analyses of the particular field. The overview specifies the challenges the programme has to address. Every programme and subprogramme in the Széchenyi Plan searches for solutions and determines objectives on the basis of the overview:

- the most important objective of the enterprise development programme is to strengthen and improve the competitiveness of already existing ventures rather than increase the absolute number of enterprises;
- the housing programme aims to boost the supply of housing and improve mobility;

- the tourism development programme focuses attention on measures necessary to assure quality tourism;
- the programme supporting research, development and innovation tries to establish a link between R&D and enterprises;
- the information society and information economy development programme aims at creating the framework of the so-called "New Economy";
- the expressway development programme aims to open up the markets of under-developed regions and improve access;
- the fundamental objective of the regional economy development programme is to develop regional economic networks and innovation.

The Széchenyi Plan relies on resources other than taxation for funding. By far the largest source of finance is provided by the business sector, and all other sources can be seen as co-financing measures in relation to this. A second source of funding is state co-financing in the form of specified appropriations in the 2001-2002 budgets; and a third source is the European Union pre-accession fund, employable for such purposes. A fourth resource available for the implementation of the Széchenyi Plan is three-pronged: a modern institutional system of state administration, subsidy systems following business models, and the meeting of requirements for international programme financing. This increases efficiency in the use of public funds.

In accordance with the planning and programming specifications of the European Union, the Széchenyi Plan foresees medium-term programmes covering a period of six years. These programmes already include a specified amount of state co-financing as regulated in the Budget Act for 2001-2002. In the period 2003 to 2006 a specific share of the annual GDP is to be set aside as a source of government co-financing for the plan. The exact figures will be specified in the budget for that particular period. The share of Government co-finance varies between programmes. As seen in Table 3 the actual amounts allocated to the Széchenyi Plan are the following: 295.9 billion HUF (1.0 billion US$) in 2001 and 330.8 billion HUF (1.2 billion US$) in 2002 (1 USD = 283 HUF, February 2001).

The amount set aside by the Hungarian government for co-financing the Széchenyi Plan in 2001 and 2002 (about 2 billion US$) is significant, and compares well with other international support programmes; it may indeed prove sufficient at this initial stage. Central co-financing opens up opportunities for businesses and local governments to implement their projects, provided these projects fit the programmes of the Széchenyi Plan. All parties will have a hand in the plan's success or failure, whether the participants of business, domestic and foreign entrepreneurs, businessmen, or the individual regions and municipalities.

Table 3. Government co-financing for the Széchenyi Plan in 2001 and 2002

Programmes	2001	2002	2002
	HUF billion		%
Enterprise development programme	31.4	37.3	10.9
Housing programme	69.9	72.6	22.8
Tourism development programme	25.0	28.1	8.5
Programme for the support of research, development and innovation	17.5	37.0	8.7
Information society and information economy development programme	15.0	28.9	7.0
Expressway development programme and related subprogrammes for infrastructure development	132.1	120.9	40.4
Regional economy development programme	5.0	6.0	1.6
TOTAL	295.9	330.0	100.0

Source: National Statistical Office, Budapest, 2002

3.3 Institutional Aspects

Decentralisation is one of the basic principles in the renewal of Hungarian spatial policy, and accords fully with the economic doctrine of the European Union. Thus 1996's Act XXI on Regional Development and Regional Planning began the specification of an institutional system of modern regional development at the lowest level. The smallest spatial category specified and governed by the regional development act is the *sub-region*, and this has in fact no administrative equivalent. Sub-regions are the smallest units eligible to receive spatial development funds and they form the local institutional framework for development activity. Support is provided mainly for settlement and sub-regional economic and infrastructure development, based on the co-operation of local governments.

The *county*, a territorial level that is also important from the administrative perspective, is given a key role in regional development. By the 30th of June 1996 county development councils had been created in the 19 counties of Hungary, facilitating the decentralisation of decision-making in regional development. County development councils organise and integrate the various regional development activities of local governments and state administrative organisations, co-ordinating operations, determining county regional plans and the distribution of decentralised governmental development resources.

In Hungary the traditional counties, when viewed as administrative units, correspond to the so-called NUTS III spatial level in the European system. However a condition for receiving regional development support from the EU is the elaboration of development programmes for much larger regions, i.e. NUTS II level. Therefore six statistical *planning regions* have been established by parliamentary approval of the National Regional Development Plan. These further evolved in 1997 to include more counties, and now cover the whole of Hungary. In the meantime new regional development councils have also been formed. These draw up and vote on regional development plans, organise and co-ordinate the regional development process and evaluate the physical and spatial development plans of the county. The forming of regions is a lengthy process, involving complex issues; it demands, among other things, more precise controls and the provision of operational conditions to determine labour methods.

The modernisation of regional policy in Hungary and the realisation of a new institutional system as specified in Act XXI./1996 has led to the creation of a new higher-level and separate central state administrative apparatus: the National Regional Development Council. This enhances the previous system for regional and spatial development. The *National Regional Development Council* is the governmental body for regional and spatial development. It provides Parliament and the government with necessary information for intelligent decision-making, harmonises national and regional development actions, offers suggestions for governmental decrees, and is a participant in the administration of spatial organisations.

The main funding for governmental support of regional, county and sub-regional development programmes is specified in two national directives, the *Regional Development Support Directive* and the *Spatial Equalisation Financial Assistance Directive*, with monies being disbursed by local government. However this does not amount to more than 5-6 % of the investment resources for development. Total investment from 1992 to 1995 was 19.14 % of GDP for this four-year period. About 35 % of all investments – 6.7 % of the GDP – came from governmental resources. Of this, 21 % was investment of local and central governmental institutions, 9.4 % was governmental support for housing construction, and 4.6 % governmental support for investment of economic organisations.

3.4 Environmental Aspects

The necessity of protecting and improving the environment was recognized both before and after the social, economic and political changes in 1989/1990. This recognition was given expression in appropriate legislation, but enforcement of laws has proved ineffective. Some progress has been made in nature conservation, with the designation of National Parks and areas of special protection, etc. Much less progress has been made in environmental improvement. Although the mining sector is obliged to rehabilitate derelict mines this obligation is only sporadically observed, and quarries, in particular, have not been rehabilitated. Thus the number of scars on the landscape is actually growing. In recent years one positive trend has been urban renewal, especially the rehabilitation and upgrading of run-down inner-city areas. National and regional programmes promote the environmental improvement of former sites of heavy industry. Most of these areas are located in an attractive natural environment and are consequently particular blots on the landscape. Environmental improvement is an important precondition of stable socioeconomic growth and development.

The National Spatial Plan identifies zones where the implementation of complex environmental improvement is urgently necessary. These zones cover continuous areas ruined by mining and related industrial activities (mountains of rubble, slag dumps, sludge reservoirs). The municipalities incorporating such zones must redevelop the areas designated for complex landscape rehabilitation in accordance with the landscape plan in the National Spatial Plan. This landscape plan is based on new objectives and aims to foster spatial interests, as specified in the spatial plans of special regions and counties. The Spatial Plans of the counties further refine the guidelines of the National Plan, and identify additional zones of regional importance for rehabilitation, upgrading and/or special protection. Although plans and regulations have been drawn up, their enforcement is yet to be achieved.

4 Conclusion

As in other countries, Hungary's old regional centres of heavy industry have suffered most drastically from the impact of economic restructuring. In the first years of the 1990s, governmental assistance for these regions had only slight mitigating effects, postponing change rather than actually bringing solutions. Central and western regions were the first to benefit from the gradual recovery of the national economy, while areas in the north-east, with high concentrations of industry, fell behind in the process of revitalisation. Only in recent years has domestic and foreign capital begun to flow into eastern and southern regions and communities, contributing to their revival. Incoming firms must be encouraged to remain and expand, thus providing employment and local income. Progress was made early on in education and training, enabling those workers willing to relocate to find employment. Several sophisticated programmes have been elaborated for the revitalisation of former heavy industrial regions, but implementation has been much slower than originally expected. A more consistent commitment is shown by the recently introduced Széchenyi Plan, and the development of infrastructure (initially of transport and water management) has already brought results.

References

Hungarian Government (2001) Széchenyi Plan, Government Resolution

Hungarian Parliament (1998) National Spatial Development Strategy, Parliamentary Resolution No. 35/1998 (III.20.), Budapest

National Spatial Plan (2001) Draft act to be presented before the Parliament

Processes of Regional Change (2001) Report on the implementation of the National Spatial Development Strategy, Budapest

Yearbooks of the National Statistical Office

Latvia: A Centre-Oriented Country in Transition

Zaiga Krišjane

University of Latvia, Human Geography Department, Riga

1 The General Situation of Industrialised Cities and Regions in Latvia

1.1 Historical Development

Although the earliest industrial ventures in Latvia appeared in the Duchy of Courland in the sixteenth century, major industrial development really began in the nineteenth century, when Latvia was a province of the Russian Empire. The location of industries was dictated by the large domestic market of Russia; Latvia was in a favourable position regarding transportation and geography. When Russia increased the volume of exported goods passing through Riga, Liepaja, Ventspils (see map) and other port cities, industrial sectors were also stimulated. Another factor driving Latvia's economy was the construction of railroad lines from the 1870s. The country became an important centre for the production of machinery, rubber, chemicals, textiles and wood. Companies specialised in mechanical engineering were also prominent, manufacturing ships, railcars, motors, steam engines, boilers, and various kinds of machine tools. In the early twentieth century the main industries in Latvia were metallurgy, machine building, the production of electrical equipment, the chemicals industry, the textile industry and the wood-processing industry. A relatively small number of large companies (about 30 % of enterprises), often employing more than 100 people, provided work for 80 % of the country's industrial employees, representing approximately 79 % of industrial output. Almost three-quarters of industrial companies were located in Riga, with about 12 % in Liepaja (Aizsilnieks 1968). Before World War I Riga was a leading industrial centre in the Russian Empire, alongside St. Petersburg, Moscow, and Warsaw. The industry concentrated in Riga produced approximately two-thirds of Latvia's industrial output (Jankevics et al. 1975).

Map: Latvia – Cities, roads and regions

At the beginning of the twentieth century factories in Riga were responsible for approximately 5 % of the Russian Empire's industrial output, even though Latvia represented just 1.5 % of the population. Latvia's factories used raw materials and preliminary products from western Europe (approximately 65 % of all materials), while finished products were almost exclusively sold in Russia; only 7 % of products were exported (Aizsilnieks 1968).

Liepaja's emergence as the second largest industrial centre in Latvia was stimulated by sectors such as steel smelting, machine building, wood-processing, textiles, shoe, and other enterprises. At the time of these developments Daugavpils and Rezekne in the eastern Latvian region of Latgale were mostly known as important trade centres.

In 1913 industry played an important role in Latvia's economic life, contributing 52 % of the province's total economic output. Moreover, capital goods such as machine tools and equipment made up a significant share of output. Industrial development facilitated urban development, in turn swelling the population; by 1914 Riga had 520,000 residents (Jankevics et al. 1975). This economic growth was also boosted by scientific developments and discoveries. The Riga Polytechnic Institute produced highly-qualified engineers and technicians and played an important role in industrial research. The existence of a well-trained labour force in Latvia is confirmed by the fact that the Russo-Balt factory in Riga produced the first automobile in the Russian empire. World War I had a devastating effect on economic development. Most industrial equipment was transported to Russia and was not returned after the war.

During the period of Latvia's independence (1918-1940) the country's economy underwent considerable restructuring. Government policy aimed at satisfying consumer demand and Latvia's domestic market absorbed approximately 80 % of industrial output. Small and medium-sized businesses were promoted by easy credit. Agrarian reform in 1920 boosted the importance of agriculture to the Latvian economy, and agro-industrial sectors were similarly stimulated. New industrial enterprises were established to compensate for population and equipment losses suffered during the war. Industry used local raw materials and focused initially on the domestic market. Metallurgy and the chemicals industry were not restored to their pre-war importance as the requisite massive capital investment could not be found. Instead textiles, food processing, metal processing, machine building, and chemicals only gradually re-emerged as the country's leading industrial sectors. Forestry, wood processing and paper production all expanded rapidly, both to satisfy domestic demand and for export (Aizsilnieks 1968). Even if industrial output did not return to the levels of 1914, Latvia could still boast of a highly talented and qualified workforce. The national electronic engineering company VEF, located in Riga, manufactured airplanes, communication equipment and the world's first Minox mini-camera; other fields in electronics and radio were also developed. As had been the case before the war, industry was concentrated in Riga. In 1938 the capital's factories were responsible for 73.2 % of all industrial output and employed 69.7 % of industrial workers. Liepaja produced 8.0 % of output, with 7.5 % of workers. Other smaller industrial centres in Latvia were Jelgava, Daugavpils, Ventspils, Kuldiga, Valmiera, and Sloka (Jankevics et al. 1975).

Latvia's annexation by the Soviet Union after World War II made industrial development a forced issue. Moscow saw industrial expansion as a way to promote russification in the Baltic States, and sent numerous workers from other parts of the Union of Socialist Soviet Republics (USSR) to Latvia to work in newly built factories. In the 1940s and 1950s industrial development expanded very quickly under major Soviet capital investments, outstripping local resources and labour. Existing industrial enterprises were expanded, such as the Riga shipbuilding and refitting works and the "Red Star" bicycle factory, with new factories to produce automobile electronics, experimental meteorological equipment and diesel engines being built in Riga, Daugavpils and Jekabpils. Researchers calculated that by 1948 Latvia's industrial output had finally returned to 1913 levels, and by 1950 the government had built or renovated more than nine hundred industrial enterprises, with output three times the 1940 level (Jankevics et al. 1975).

These data illustrate the very rapid pace of post-war industrialization. Ideological considerations, though dominant, were in this case sympathetic to Latvia's tradition of industry and qualified labour force, its relatively well-developed infrastructure and outstanding geographic location. Under the principle of specialisation, whereby each Soviet republic was accorded a specific industrial sector, the Soviet authorities designated machine building and metal processing to be the leading industrial sectors in Latvia, focusing on the production of electric engines, diesel engines, electric railroad cars, electronics for automobiles, vans, and other items. The machine building sector, requiring large amounts metals and energy, was greatly expanded.

Industry in Latvia was developed in accordance with the political decisions of the Communist Party of the Soviet Union (CPSU), the five-year plans of the USSR and the deliberations of the Soviet Union's sector-based ministries. Decisions to build factories were often issued in the form of directives. In the 1950s and 1960s the building of new enterprises and the expansion of older ones continued – VEF (a manufacturer of electro-technical equipment) and the Riga Carriage Building Plant were both expanded. New factories were built outside the cities, providing employment for entire worker-towns; some of theses consolidated into permanent cities over the ensuing decades. Such so-called "new cities" are Olaine, Salaspils, Aizkraukle, and Vangazi. Between 1960 and 1980 major factories were built in Daugavpils, Rezekne, Valmiera, Olaine, and Ventspils. Rapid industrialization also occurred in eastern Latvia, not previously associated with large-scale industry. Latvia's second largest city, Daugavpils, became an important centre for the production of machinery and chemical fibres; Rezekne became the site for equipment-building factories and several food-processing companies.

The chemicals industry was seen as one of the leading sectors in what the Soviet Union called the "scientific and technical revolution". Built in the 1960s, the new-town of Olaine was planned as a centre for industrial chemistry, with factories that would manufacture chemical products being designed by Latvia's scientific research institutes. Similar examples were seen in the development of Aizkraukle and Salaspils, both connected to the construction of hydroelectric plants on the Daugava River.

The building-materials industry developed extensively in Latvia, stimulated by demand for construction materials for industrial projects, public and residential buildings; this benefited larger towns and smaller settlements such as Vangazi, Garkalne, and Sauriesi. Vast monoliths of identical apartment buildings were erected from the mid-1960s, especially in larger cities where workers could settle close to the new factories. Affiliates of major factories were placed in smaller towns or rural centres. By the early 1980s machine building and metal processing had emerged as the leading industrial sectors in Latvia, employing 38.4 % of all industrial workers; light industry employed 20.5 %; the food industry 12 %; forestry, wood processing, cellulose and paper manufacturing 8.8 %; and the chemical industry 5.9 % (Rupnieciba 1984). Although the Council of Ministers of the Latvian Socialist Soviet Republic (SSR) issued a decree that industry should be decentralised over the country, still the majority of Latvian enterprises continued to be concentrated in and around the capital city, Riga. This led to increased environmental damage; until the late 1980s industry was the major source of air and water pollution in Latvia.

The centralized decisions on industrial development in the Soviet Union created an artificial shortage of employees in Latvia, leading to a large inflow of workers from other regions of the USSR. Between 1946 and 1989 about 700,000 people moved to the Latvian SSR from other parts of the Soviet Union, approximately 400,000 of them after 1960. The migration quickly outstripped natural population growth. Most of the immigrants came from Russia, Belarus and the Ukraine, so that the ethnic structure of Latvia underwent drastic change (Migracija 1998); in addition the elimination of individual family farms and forced collectivization of agriculture led many rural residents to depart for the cities where wages were higher.

At the end of the 1980s, shortly before the Soviet Union began to disintegrate, the following industrial sectors were important to the Latvian economy:

- sectors based on imported raw materials: machine building, metal processing, chemicals and light industries (Riga, Daugavpils, Rezekne, Jelgava, Valmiera, Olaine, Jekabpils, etc.);
- intersectoral complexes based on local raw materials: forestry-related industries, including furniture-making and cellulose/paper industries; construction materials; maritime industries such as fishing, fish processing, sea transport and harbour management (Ventspils, Riga, Liepaja, Salacgriva, Kuldiga, Broceni, Jurmala, etc.);
- agricultural industries: agriculture and agricultural processing (Riga, Daugavpils, Rezekne, Jelgava, Valmiera, Bauska, Sabile, Aizpute, etc.) (Jankevics 1991).

In the 1980s Latvia ranked sixth among the Soviet Union's fifteen republics in terms of industrial output, even though in terms of population it ranked fourteenth. Aside from industrial development it is worth mentioning that the country's outstanding beaches and medicinal mud baths and streams helped promote the development of the spa and tourism sector. Jurmala was a prized holiday destination within the USSR; huge sanatoriums and rest homes were built for workers from

Soviet ministries and labour unions. Latvia's advantageous location on the shores of the Baltic Sea also determined its specialization in the transport sector, with cargo from all over the Soviet Union being transported through Latvia's ports.

The process of industrialisation during the Soviet period had a tremendous impact on Latvia's employment structure. In 1988 40 % of the workforce was employed in industry and construction; the share in agriculture and forestry had declined by a factor of 4.5 since 1935, to just 15 %. The service industries employed only 45 % of workers (Jankevics 1991). Industry in the Soviet Union was highly concentrated; 65 % of people worked in companies with more than a thousand employees (Rupnieciba 1984). Under conditions of centralized planning, Latvia had more than four hundred independent industrial enterprises in the late 1980s; major industrial associations (88 in 1988) and scientific production associations had also begun to emerge (Jankevics 1991).

The Soviet-driven expansion of industry left Latvia in a difficult position. The nation's industrial sector was structurally deformed, as output had been dictated by economic requirements in other parts of the Soviet Union. Native industry did not make full use of local raw materials; nor did it promote the reprocessing of materials or provide agriculture with equipment necessary for local conditions. The manufacturing sector was relatively well developed, though plagued by irregular supplies of energy and heating fuel and recurring difficulties in the supply of raw materials. Many industrial enterprises in places like Riga did not have links with local firms, a consequence of the direct subordination to Soviet and republic-level ministries. Enterprises subject to the control of Soviet ministries in Moscow produced 37 % of industrial output in the Latvian SSR; many of these enterprises were closely linked to the military-industrial complex.

1.2 Present Development and Future Prospects

Industrial development is closely linked to a country's general economic development. For a better understanding of the current situation in Latvia's industrial centres it is necessary to look at the processes of transformation which occurred over the previous decade. Alongside political reforms, Latvia underwent a complicated process of economic transformation in the 1990s in order to create a market economy. This is expected to ensure a stable and sustainable economic development, as well as setting the foundations for increased welfare and security of the population.

Latvia, like many other countries in central and eastern Europe, decided to implement a fairly radical model of reforms, hoping to bring democracy and the market economy as quickly as possible. This has proved to be a painful and difficult process. The first three years of Latvia's independence (until 1993) saw the collapse of the centralised planning system, illustrated by Latvia's falling GDP. It declined very rapidly, reaching a low in 1993 of only 49.6 % of the 1990 value (CSBL 1999b, 27). This implosion of the planned economy, together with price liberalization and ensuing inflation, as well as other factors such as currency reform and a rapid loss of sales markets, led to a serious decline in economic activities in the first few years of the transformation. However GDP increased in 1994, suggesting that Latvia was beginning to recover from the crisis. The economy continued to grow in subsequent years most notably in 1997, when GDP rose by 8.6 %.

Changes in the Latvian economy were not only reflected in GDP fluctuations, but also in structural shifts. From 1990 the share of the national economy represented by agriculture fell year by year to just 3.9 % at the end of the decade. Another decisive change was the vast expansion of the service sector, now the leading sector in Latvia's economy. The manufacturing sector has diminished considerably. Many enterprises were shut down in the previous decade or have reduced output, with the machine building sector being particularly affected (see Table 1).

A shrinking industrial sector over the past few years has been compensated by rapid growth in the construction industry and other service-related sectors. Transport and communication have become leading service sectors in Latvia, especially in the area of transit services. A decline in manufacturing output in 1999 of nearly 10 % was significant, following in the wake of the Russian financial crisis in the last two quarters of 1998 (Ministry of Economy 2000a).

Table 1. GDP – structural changes, 1990-1999 (in %)

	1990	1995	1996	1997	1998	1999
Agriculture, hunting and forestry	21.8	10.4	8.7	5.6	4.5	3.9
Fishing	0.8	0.4	0.3	0.2	0.2	0.1
Mining and quarrying	0.2	0.2	0.2	0.2	0.2	0.2
Manufacturing	34.5	22.4	20.9	22.2	20.2	14.8
Electricity, gas and water supply	1.8	5.5	5.3	5.0	3.9	5.0
Construction	9.7	5.1	4.7	4.8	5.2	7.6
Services	31.9	56.0	59.9	62.0	65.8	68.4

Source: CSBL 2000b, 24

The current situation in manufacturing industries shows differences among various sectors in terms of growth rates. This has a knock-on effect on the development of regions and towns, and the number of jobs has continued to decline in Latvia's larger cities. In 1990 more than half of workers in Liepaja were employed in the industrial sector, with 43 % in Daugavpils, 26 % in Ventspils and 20 % in Jurmala. In 1998, by contrast, only 9.5 % of workers in the harbour city of Ventspils were employed in manufacturing, though higher levels persisted in Daugavpils (28.6 %) and Liepaja (27.9 %) (based on statistical data, CSBL 2000a).

An analysis of the structures of employment in Latvia's larger cities shows that structural change has had less impact on those cities such as Riga which have a diverse industrial base. Despite the fact that Riga and its metropolitan area have traditionally had a high concentration of industrial enterprises, only 40 % of the city's workforce were employed in industry in 1990; today the figure has slipped to below 20 %. Small specialized industrial centres are facing an even worse situation, with many enterprises either closing or being forced to run-down production considerably. It should be noted that there are, of course, exceptions, and some specialized sectors have actually expanded in certain regions. The wood-processing industry supports many small rural centres, while the fishing industry is the main source of employment in coastal regions.

The *food-processing industry* has now emerged as the most important sector regarding value added. Enterprises engaged in food-processing are found in virtually all of Latvia's district centres, as well as in most small towns. If prior to 1990 the food-processing industry was dominated by large firms processing meat and dairy products (as well as other agricultural produce), today there are many small and medium-sized businesses in the mix.

Light industry has a long history in Latvia; traditionally it developed in the country's largest cities, with some companies or affiliates in smaller towns. The largest company in the *textile industry* in Latvia is a clothing factory in Ogre, one of the largest in eastern Europe in terms of output capacity. Textile products are Latvia's second largest field of export, surpassing even foodstuffs. However, following the Russian economic crisis the export of textiles to Russia and the rest of the Commonwealth of Independent States (CIS) has fallen by half; today 80 % of Latvia's textile articles are exported to the member states of the European Union. The textile industry is labour intensive, and labour costs can therefore be an important factor regarding international competitiveness. The training of employees is not expensive, and the sector requires relatively little in the way of capital investment. Thus transitional countries are often able to compete effectively in Western markets with their cheaper textile products, especially when joint venture operations are established (Ministry of Economy 2000a).

The *wood-processing industry* has a stable source of raw materials (approximately 45 % of Latvia is forested) and is the most dynamic of Latvia's processing industries. It can boast highly qualified workers and healthy sales markets (mostly in the EU). The industry has excellent opportunities of increasing its income levels by exporting more fully processed wood. Wood materials are now Latvia's leading export and 94 % of the industry's output is exported. The largest companies in this sector are located in the cities of Riga and Ventspils, as well as the districts of Riga, Tukums, Cesis, Kuldiga, and Madona.

The *chemicals industry* has a long and stable history in Latvia, shown today by the good system of scientific research in the field of chemistry and the highly-qualified workforce. Major industry centres are Riga, Daugavpils, Olaine, and Dobele. Companies in these towns make a wide variety of chemical products – artificial fibres, plastics, paints and lacquers, medicines, perfumes, cosmetics, etc. The main export markets for the chemical industry are the countries of the CIS. Russia's economic crisis led to a considerable narrowing of the sector in 1999.

Another industrial sector based on local and natural raw materials is the *building-materials industry*. It uses materials such as clay, sand, gravel, limestone and gypsum; the main areas of activity are in the cities of Riga, Broceni, Jelgava, and Kalnciems, and the districts of Limbazi, Preili, and Cesis.

As already mentioned the proportion of Latvia's industrial sector represented by the *machine-building industry* has declined significantly in the last few years.

Household electrical appliances, radios and motor vehicles are rarely manufactured now in Latvia, and production of electric trains and tram cars has stopped. This industry is one in which transformations have been particularly negative; one reason for this is that a competitive machine building industry requires massive capital investment (Ministry of Economy 2000a).

Any country that wishes to develop its industry and to attract investments must develop a favourable and stable economic climate. Certain factors are decisive, such as the distribution of natural resources, the quality of available labour and an attractive business environment.

Latvia's transport and communications sector has been the most attractive field for foreign investors up to now. Latvia has once again become an important transportation and transit hub. The sectors woodworking and food processing have received the bulk of investment. The most promising production sectors for foreign investment are information technologies, electrical and mechanical engineering, chemical and pharmaceutical industries, wood processing, food processing and textiles. An excellent transport infrastructure, a long industrial tradition and western-style work practices support investment in all of the aforementioned sectors (Latvian Development Agency 2000).

Today output is increasing mainly in labour intensive industries with a low value added, and cheap workers are being attracted to fill these new jobs. However, Latvia's long-term interests lie in scientifically and technologically intensive sectors with a high value added and export potential. In the future, increases in industrial output will be driven primarily by expansions in such exports (Kage 2000).

2 Specific Problems in Industrialised Cities and Regions in Latvia

Latvia and other countries in central and eastern Europe were subject to centralized planning for decades and show similar problems during the process of transformation. Of course the considerable differences between these countries has dictated the pace and degree of economic and societal development. Historically Latvia has had an extensive infrastructure, qualified and educated workers, well-developed industry and agriculture, and an outstanding geographic location. These conditions, along with recent experiences in democracy, traditions in business and distinct national self-awareness give Latvia certain advantages in the transition process.

Latvia was completely integrated into the Soviet economic and political system, unlike some other countries in Central Europe (Dreifelds 1996, 110). Accordingly it was the scene of a much more thorough and all-encompassing economic restructuring. After the collapse of the Soviet Union (SU) Latvia did not only have to rid itself of the more onerous economic restraints, but also had to create the economic tools necessary for a modern independent country: a tax system, a national budget and customs institutions, amongst other things. During Soviet times the Latvian economy was regulated and controlled by Moscow, with the implementation of central plans overseen by sector-based ministries. Each of these had complete control over the development of its sector, and, as detailed above, Latvia's specific conditions were all but ignored in the process. When the system collapsed Latvia lost most of its export market. This was an economic disaster, particularly for the industrial regions. Over decades production intended for the markets of the SU and the COMECON had deteriorated in quality to such an extent that many Latvian products subsequently proved highly uncompetitive in developed markets. In the course of transformation to a market economy many companies were forced to shut down, or at least to restructure and modernise to meet market requirements.

Another key issue for the Latvian government was how best to restructure the national economy. Centralized planning of the Soviet era had bequeathed a collection of unprofitable, polluting, and obsolete industrial enterprises, without access to raw materials or a viable sales market; none could compete on the international market. As in other countries with a planned economy, Latvia's industrial sector had been excessively promoted at the expense of services. All of these problems have had a hand in creating the current social difficulties. Latvia was also forced to adapt to the international economy, an economy that was rapidly changing under the pressures of globalisation. Latvian society had to face many challenges in adapting to new social, political, and economic conditions. New forms of thinking and behaviour gradually emerged, though not everyone was willing or able to make the necessary changes. Although the collapse of the Communist regime and the rebirth of democracy and a market economy radically expanded people's social, economic, political, and cultural opportunities, the problems of alienation, stratification and financial insecurity also burdened many individuals (Kage 2000, UNDP 1998).

The standard of living in Latvia plummeted for a majority of the population at the beginning of economic reform, with both income and savings showing a decline. Inequality has increased, and there are now indications that certain sectors of the population are living in poverty (UNDP 1997, 1998).

The restructuring of the economy since 1990 has brought about major changes in the employment structure. The number of people employed in the national economy has been in continual decline since the beginning of the transformation process (with the exception of 1997). In 1990 1.4 million people had jobs; in 1999 the figure was around 1 million, out of about 2.4 million inhabitants. (CSBL 2000b) Changes in the structure of the economy led to changes in the capacity of various sectors to provide employment. The number of workers in industry, construction and agriculture declined severely; increased employment levels were seen in the public sector (government), defence, social welfare, financial mediation, and a few other sectors. In 1998 the number of people working in the service sector considerably exceeded the number of those working in production. Increasing unemployment has resulted from the rapid reorganisation and privatisation of companies. The increase in the number of jobs in the information technology sector is highly significant, with specialists being in particular demand. Latvia began registering unemployment figures in 1992, with the number of jobless people increasing constantly since then (CSBL 1998, 52). In 1998 the official unemployment rate, measured as the ratio of the registered unemployed to economically active residents, was 7.6 %. According to ILO the unemployment rate was 14.0 % (Ministry of Economy 1999, 59-60).

Official unemployment in Latvia varies considerably from region to region, with some areas showing three times the national average; eastern Latvia is worst affected, with an unemployment rate of 20 %. In this mostly rural area many large industrial enterprises have simply not been able to adapt to the new economic conditions. In many cases these were companies that were previously oriented towards markets in Russia and the CIS. In Riga and Riga district, by contrast, unemployment is comparatively low; and it is even lower in the port city of Ventspils, where a large harbour system on the eastern shore of the Baltic Sea serves to maintain employment.

Privatisation is one of the typical economic processes implemented in a country in transition. This helps to develop the business environment of the country, as well as accelerating the processes of change. In Latvia, as elsewhere, the purpose of privatisation is to reduce the state's role in business and to create a favourable environment for private capital, thus attracting additional investment into the economy. Privatisation measures should be applied to all businesses in a market economy and not just to the industrial centres. In Latvia the vast majority of state-owned companies and enterprises (97 %) have been turned over for privatisation, and more than 1,500 privatisation agreements have been concluded. The proportion of the private sector in terms of value added rose to 66 % by the beginning of 2000, with the private sector employing 70 % of the national workforce. Privatisation in Latvia has been almost completed in the extracting and processing industries, the construction industry, gas, fishing, agriculture, hunting, and forestry. In the extracting and processing industries, the construction industry, and the wholesale and retail trade sectors more than 95 % of value added is already produced by private companies (Ministry of Economy 2000a, 102). The lowest level of privatisation has been in the areas of electricity, gas, and water supply; as of writing it seems that the Latvian electricity company will remain public for the time being. The proportion of employed persons working in private companies at the end of 1998 was 57 % (these calculations, it must be said, do not take into account those who work on farms and self-employed persons). In larger cities more than half of workers have private-sector jobs. The highest proportions are found in Riga, Liepaja, Ventspils and the Riga District. Less than 40 % of jobs are private sector in eastern Latvia (the districts of Balvi, Ludza, and Rezekne), where there is also a high level of unemployment (CSBL 1999a). Small and medium-sized companies are a key element in a market economy, and today more than 65 % of all jobs are in such companies.

The country's unbalanced spatial development and the differences in regional economic activity can be described by GDPs and a sectoral analysis. The highest income levels are found in regions with flourishing exports and where the restructuring of economic sectors has been realised most successfully. The per capita GDP figures analysed by region (the statistical regions on NUTS 3 level, see map) show the disparities between the economically powerful capital city and its metropolitan area, producing more than 60 % of industrial output and market services, and Latgale, which contributes less than 10 % of the country's GDP (see Table 2).

Table 2. Gross domestic product by region, 1997

Region (NUTS 3)	GDP (%)	regional share for each sector of value added (%)			GDP per capita, in lats (Ls)
		agricul-ture	industry	services	
Total	100.0	100.0	100.0	100.0	1,327
Riga region	56.1	20.0	60.7	62.9	1,810
of which in Riga	48.9	4.5	52.2	58.1	1,975
Vidzeme region	9.4	25.0	10.0	6.1	842
Kurzeme region	15.4	16.3	11.9	15.8	1,499
Zemgale region	9.1	26.6	8.3	6.1	840
Latgale region	9.9	12.0	9.2	9.1	820

Source: Ministry of Economy 2000a, 20

When the GDP per capita is compared at the level of regional territorial units (the NUTS 4 level) even greater variations are seen: the highest indicator is 4.4 times the lowest indicator. The income level in the city of Ventspils, for example, exceeds the national average by 3.75 % (CSBL 2000a). The main sources of income come from transit, especially the export of Russian oil products through the port. The lowest indicators are found in eastern Latvia, where the GDP per capita is only one third the national average. This is due to a combination of factors: the low level of economic activity in rural areas, structural problems in the region, and a lack of skills and knowledge on the part of locals in small and medium-sized towns on how to promote business activity.

One problem common to industrial and non-industrial regions in Latvia is the uneven distribution of investment. Riga and its adjacencies are the main beneficiaries of investment with more than 60 % of all foreign investment. The port cities of Ventspils and Liepaja are the next most popular foreign investment locations. Much of the state's national investment programme in 1999 was targeted toward the country's larger cities: 17.1 % for Riga, 12 % for Ventspils, 3.9 % for Liepaja, 3.2 % for Daugavpils, and 1.3 % for Jelgava. By contrast, 38.5 % of programme resources were devoted to smaller projects in other Latvian districts, as well as going towards projects which cover all of Latvia or a considerable share of the country. The priorities in the investment programme include energy, transportation, environmental protection and public welfare (Ministry of Economy 2000a).

One problem specific to industrial centres is how to utilise old factories and land that was the site of industrial activity. Often new projects are built on unimproved land, frequently at the edge of cities or in rural areas.

3 Strategies for Further Development

3.1 Discussion and Studies on Regional Development

According to various socio-economic indicators Latvia shows strong regional differences which are exacerbated by the dominance of the city of Riga. These interior disparities must be closely examined in order to encourage an equalisation of territorial development; this will be achieved through the visions contained in political and policy documents and appropriate measures based on scientific research. Traditionally historic regions of Latvia – Kurzeme, Vidzeme, Zemgale and Latgale – are considered. Contemporary regionalisation processes are influenced by eight economic regions that were formed in the 1970s and 1980s. They are supposed to provide equalised economic territorial development and decrease the concentration of production branches in Riga by building new enterprises in regional centres. Analysts of this government policy have sometimes stressed that it has, in fact, been the cause of important social and economic problems in those centres, including a rapid growth of unemployment and decreased industrial production (UNDP 1997). The issues of regional development and politics are wide and complex, impinging on many theoretical disciplines; in Latvia many academics and researchers from diverse branches have made it their focus for study. Some aspects have already been carefully analysed - for example the role of border areas, the question of cities in regional development, aspects of populations dynamics and human capital development, innovation and the role of tourism in regional development (Melluma 2000, Krišjane and Jankevics 2000, Eglite 2000b, Kruzmetra 2000, Nikodemus and Rozite 2000). Many studies also assess transformation processes and outline different solutions for improving the current situation, with prognoses given. The other broad topic in regional policy and development is EU enlargement (Karnite 1999). Since the 1990s, discussions on regional development have been closely connected to various reform processes in the country. The most essential of these were administration reform (i.e. changes to the structure and functions of the state), the role of local governments, and processes of administrative territorial reform.

These reforms were initiated at the beginning of the 1990s and are ongoing (Skinkis and Steins 1992). Larger local and regional municipalities were originally planned, with stress placed on the necessity of decentralising many administrative activities. In the following years the economic bases of municipalities were assessed and certain models for regional administration discussed. In 2001 the debate was reanimated by the question of which of two territorial (local) level division alternatives to implement (33 or 102 administrative units). The most important aspects in this area are as follows:

- the various responsibilities and functions of local governments at different levels; the division between local and regional level municipalities (Pocs 1999);
- the Municipal Equalisation Fund and the mechanisms of its financing (Karnite 1996);
- the formation of competitive regions (Keiss and Kazinovskis 2001);

- the various models of territorial and regional division and their analysis (Rivza 1999a, 1999b);
- estimates of the adequate size of territorial units (division) for the needs of administrative-territorial reform, analysing the possible options of merging local governments and improving co-operation procedures.

The project concerning the State Administrative Regional Division (MoEPRD 1998b), prepared in 1998, should be mentioned as the best known and the most widely discussed research in this field. It analyses the administrative territorial division in regard to its previous forms and the current situation, and assesses the national administrative functions required of such territorial division. The various factors that influence administrative territorial division are outlined and analysed. Several research institutes participated in the work. A team from the Department of Geography and Earth Sciences at the University of Latvia, under the direction of P. Skinkis, studied the historical development of the current administrative territorial division of Latvia. Researchers from the Institute of Economics at the Academy of Sciences, under the direction of J. Pocs, analysed the system and structures of state administration and the division of functions among different administrative levels; and scientists from the Agricultural University of Latvia, under the leadership of B. Rivza, determined and analysed the factors that influence administrative territorial division in Latvia. Several publications are connected with this project (Pocs 1999, Rivza et. al 1999). Alternative regional divisions have been described using different measurements and methods, with eight proposed regional development options being compared and analysed and a concrete solution offered. One current area of research is the analysis of different measurements, involving the selection of indicators to describe different regions or territories and the processes taking place there, including measurements that track development (Eglite and Markausa 1993, Vanags and Vilka 2000). Although the range of measurements for analysis is wide, indicators in the following spheres can be adopted:

- identification of territories which should be specially supported (Melluma 1996);
- the parameters of economic and structural processes;
- human and social capital estimates (Eglite 2000a, Kruzmetra 2000);
- the standard of living and quality-of-life of inhabitants (Krišjane 1998, CSBL 1996, 2001);
- investment climate and environment (Spica 2000).

A clear consensus on the direction of regional policy in Latvia does not exist and these issues are the subject of broad discussion. The drafting of a National Regional Development Programme and its subprograms is ongoing, with contentious areas being the co-ordination of areas of responsibility, the determining of financing instruments and work concerning regional development. Many efficient (applied) studies have been carried out to this end (Karnite 1999, Vaidere and Vanags 1997, Vaidere 1999, Skinkis 2001). The work of A. Melluma (2000) focuses on border areas and gives an analysis of development possibilities. Studies on rural development problems have also been carried out (Boruks et al. 2000, Luse and Grave 1993). Regional development problems have been described and analysed in connection with spatial development planning. A good deal of research has been undertaken in this area in connection with the drafting of National Planning and the assessment of the planning process itself. Along with the drafting of different level spatial development plans, helpful methodological studies and materials are being prepared, for example, the analysis of the Kuldiga district to prepare a regional territorial development plan, as well as manuals for municipality spatial development planning (MoEPRD 1999).

3.2 The Objectives of the Government's Economic Policy

Latvia does not have a specific development strategy for industrial regions; issues related to industry are contained in regional development plans for a wide variety of territorial units (cities, districts, regions, etc.), as well as in the country's Regional Development Concept. Many local governments are still drawing up their own development strategies. The concept on Latvia's Regional Development Policy also sets the initial preconditions necessary to improve environmental, living and working conditions in all the regions of Latvia, as well as aiming to decrease and prevent unfavourable regional differences and support those differences which are positive (MoEPRD 1998a). A Regional Development Fund was planned along with the introduction of legal measures for specially supported areas in order to realise the gaols mentioned above. The national support system for depressed areas was launched in 1996. Special economic incentives were created to promote economic development in regions where development had slowed down. Economically less-developed regions and those regions where economic activity was restricted by red-tape were offered the status of assisted areas after the development programme for the specific region was drafted (Ministry of Economy 2000a).

Latvia's government has set out goals for the development of specific sectors and national economic development. The country's interests and priorities in pursuit of membership in the EU have also been defined. Objectives of the government's economic policy are continuous and sustainable economic growth, balanced economic and social development, gradual elimination of regional and social inequalities as well as disparities in labour opportunities, protection of the environment, and efficient utilization of resources (Republic of Latvia/European Commission 1999). The economic policy must focus on ways of reducing unemployment. A purposeful state policy is needed, aiming at better standards of living; backward regions must be stimulated, the efficiency of the education and health care system increased and the social security system further developed (Republic of Latvia 2000).

Riga Rises to the Challenge

The capital Riga as Latvia's major commercial centre has experienced a severe decline in manufacturing employment but the growth of service employment is starting. Riga reached its highest population in 1990 with a total of 912 thousand inhabitants. Then the population has decreased to 798 thousand inhabitants in 2000 because of emigration and the death rate being higher than the birth rate.

Rising to the challenge of transformation the municipality has changed its political and administrative structure as well as the division of responsibilities. It has started to plan for the future. In 1993 Riga started to work on the first post-socialist City Master Plan in Latvia, which was finished in 1996 and later adopted by the municipality. To avoid chaotic decision-making in relation to city development, the City additionally started a strategic planning and management programme and the preparation of the City Development Strategy.

The city of Riga is at present preparing and soon will adopt a comprehensive Economic Development Strategy. In order to answer questions related to the City's future development, a series of research projects was undertaken. A major emphasis was put on the impact of current economic restructuring on Riga's spatial structure. Concerning the industrial areas of the city the main findings are the following.

Firstly, large parts of the city's industrial structure will remain underutilised. Many of the existing production sites and buildings are unsuited to modern production methods. Although areas devoted to manufacturing may remain significant, fewer employees will be employed in each plant. Rail transport in Riga will be replaced by greater reliance on truck transport for many manufactured goods. Secondly, opportunities will need to be found to convert industrial land to other uses. To facilitate the change, conversion will need to be made to a higher value land use. These uses tend to be retail, other commercial and residential. Thirdly, new industrial operations will want to locate on the city's periphery. While vacant older industrial space will exist, most new manufacturers will want to develop new modern plants. New vacant building sites will need to be found for these purposes.

In Latvia the amount of the real estate or property tax is paid based on the cadastral value of land. These values are set by the State Land register. The author's research indicates that the cadastral value of land for industrial purposes is too high to encourage any developers "recycle" the enclaves of partially used or abandoned industrial buildings or sites (Francis 2003). Thus, in order to start the regeneration of the industrial zone of Riga, the municipality should first alter the existing cadastral values and prepare a package of other incentives that would make the older sites suitable to any new developments. This task becomes extremely important now that the City is preparing for the new development plan of Riga 2006-2018.

Ilgvars Francis, City Development Department, Riga City Council

Industrial policy in Latvia is based on economic development priorities as set out in an agreement concerning the development of relations between the European Union and its member states (on the one hand) and the Republic of Latvia. Economic policy priorities, Latvia's middle-term economic strategies related to European Union accession and the goals of the national programme are represented in the Joint Assessment of Economic Policy Priorities of Latvia and in the Strategy for the Integration into the European Union. Latvia is continuing to pursue an industrial policy which is primarily focused on the development of an open and competitive market. Various measures have already been undertaken to restructure the industrial sector. The goal of an industrial strategy from 1999 was the formulation of a middle-term industrial policy to ensure the competitiveness of Latvia's businesses on the EU market. The priorities at the current phase of industrial development are to create favourable conditions for the modernisation and restructuring of production processes, the expansion of exports, and the even development of industry throughout the country. The industrial strategy is based on the following principles:

- stable macroeconomic conditions (fiscal stability and stable currency exchange policies) and a movement toward a fully functioning market economy;
- the establishment of a sensible legislation framework to regulate the business environment, improvement of tax policy, reduction of administrative burdens, and other steps to ensure ongoing economic growth;
- strengthening of effective competition;
- support for small and medium-sized companies;
- strengthening of Latvia's comparative advantages by improving the quality of the labour force (better education and training), promoting the development of science-intensive technologies and innovations, taking steps to improve the quality and volume of exports, building a modern and efficient infrastructure (Ministry of Economics 2000c).

Industrial development in the middle and long-terms will depend on the promotion of innovation and other factors which increase the value added of products. Stable industrial growth will be ensured not only by income from accumulated investments, but also by the establishment of a modern infrastructure and the economic benefits of innovation (Cabinet of Ministers 2000). Furthermore, a long-term economic strategy must stress the development of an economy based on knowledge and modern technologies. Only thus can the national economy become more internationally competitive and provide a basis for high standards of living for Latvians. There are three development priorities for an economic strategy:

- the formation of favourable conditions for optimal economic functioning (attractive institutional environment for entrepreneurship, a well-functioning base infrastructure, equal and fair competition, stable macroeconomic environment);
- the stimulation of effective and competitive industry structures (a rise in productive capacity, human resources development, the formation of an economy open to innovations);

- decrease of socio-economic disparity and risk (balanced regional development, decrease of social inequity, effective system of nature conservation) (Ministry of Economy 2000b).

The equalisation of socio-economic disparities is proposed in the National Programme on the Regional Development of Latvia, currently being worked out. It includes the following sub-programmes:

- the Latvian Development Programme for border-regions;
- the Latvian Rural Development Programme;
- the Renovation Programme for Rural Areas in Latvia (improvement of the physical, social and economic environment);
- the Development of surroundings of the city centres (to ensure sustainable development, improve well-being);
- Coastal Development Programme (including the programme for the development of small ports);
- the Development Programme of Sensitive Areas (former military port areas). (Ministry of Economy 2000a, 100)

The implementation of the programme for the regional development of Latvia is closely linked to the Public Investment Programme and the establishment of planning regions; basic principles for the work of councils and development agencies must also be set out (Ministry of Economy 2000a). When a strategy for the development of industrial regions is elaborated, attention must be devoted to the goals, priorities, and action plans contained in the European Spatial Development Perspective (ESDP), Visions and Strategies around the Baltic (VASAB) 2010, and the Baltic Agenda 21. Latvia has several regional development strategies:

- a pilot development plan for the Latgale region (PRO-INTER CV et al. 2000, 5);
- a strategy for the development of the Riga region up to 2010 (Riga Region Development Council 1999);
- a spatial structure plan for the Liepaja region (Liepaja district Council et al. 2000).

Some of these plans are still being worked, and they include both bottom-up and top-down approaches. The planning documents firstly evaluate the existing situation, then define relationship priorities, outline the proposed development vision in some cases, and finally offer specific steps to improve the situation in the region.

Of key importance in territorial development is the issue of "hard" location factors, comprising primarily the types of infrastructure. Major investments from both the national investment fund and the PHARE programme are aimed at achieving environmental and infrastructure improvements. A region's accessibility depends on the state of the transportation and communication infrastructures, and development of the latter requires substantial investment. If Latvia's structurally weaker territories are given the status of regions requiring special support they receive tax discounts and other forms of support for business development from the

Regional Development Fund. "Soft" location factors are also of critical importance, such as having active local politicians and local government officials. They must be able to agree on common goals and priorities and pool their knowledge and experience on development strategies and planning. The views of residents are also important, as is their readiness to participate in the elaboration of the strategy and its implementation. Both parties, the authorities and the public, must be galvanised if a creative environment is to be established for effective development.

Below are presented two examples of how Latvia's regions see their development. Both areas were industrialized in the past, and in the outlined strategies we see that Latgale must overcome structural backwardness, while the Riga District is a territory which must compete with other leading metropolitan areas in Europe.

3.3 The Pilot Regional Development Plan for the Latgale Region

The pilot regional development plan for Latgale was drawn up by a consortium comprising PRO-INTER CV (Belgium), AGROTEC SpA (Italy), WES (Belgium) and a group of local experts. The plan was developed in 1997 under the auspices of the PHARE programme. Latgale is considered the most disadvantaged region in Latvia, and also belongs to those areas most dependent on the old economic structures developed by the Soviet economy. An endogenous bottom-up approach would, alone, be inadequate to facilitate development processes in such a backward region. Expert opinion is that exogenous top-down measures with relevant support must also be implemented. In this case the most important institutional instruments will be the Regional Council and the Latgale Regional Development Agency, being the umbrella organizations for the whole of Latgale region. They can mediate between local and district government and the national government (PRO-INTER CV et al. 2000).

The aims of the project are outlined in its general objective statement (PRO-INTER CV et al. 2000, 5): "to develop the capacity of the Latgale region in relation to regional development activities and to create an ability to effectively access national and international assistance programs, especially in the context of EU integration as well as self sustaining development of the region." Specific objectives of the project include preparation of an "integrated regional development strategy for Latgale" and provision of a comprehensive framework for short, middle, and long-term development, expected to result in the establishment of a "Latgale Regional Development Agency".

Experts have defined eight major strategic objectives. Five of these pertain to productive sectors: creation of a favourable environment for investment, development of new technologies, improvements in agricultural productivity, tourism based on the region's natural and cultural resources, and development of small and medium-sized enterprises. The other priorities are human resource development, development of communication infrastructure, and improved accessibility through better international transport corridors. (PRO-INTER CV et al. 2000).

A Regional Innovation Strategy and Action Plan should be developed as a follow-up to the pilot regional development plan. A spatial development strategy or structure plan should also be formulated for Latgale, primarily focused on preparing well-equipped industrial and multi-use sites to attract foreign direct investment. The triangle Jekabpils – Daugavpils – Rezekne should become the development nuclei and growth centre for the structurally weak region of Latgale, thus forming a city network (PRO-INTER CV et al. 2000).

The cities of Latgale suffer from various economic and social development problems. The new Latgale Urban Development Strategy, drawn up in 2000 and supplementing the Regional Strategy (Latgale Regional Development Agency 2000), aims to treat more of these diverse problems.

3.4 A Strategy for the Development of the Riga Region up to 2010

The Riga region is home to nearly half of Latvia's population and can boast a variety of economic activities. A development strategy has been drawn up for the region by the Riga Region Development Council; this strategy has great potential for producing change, particularly if diverse forces are harnessed in pursuit of common goals. The approach here is to pinpoint those development goals, priorities, and tasks that promote the development of the region and its various local governments. Contained in the strategy are:

- a brief evaluation of the existing situation;
- a vision of the potential future of the Riga region;
- the developmental directions to be taken, general and specific goals, and the tasks which must be given priority (Riga Region Development Council 1999).

The main goal of the development strategy is to create conditions for sustainable and balanced development, ensuring that the region is competitive and can integrate into Europe; thus diverse economic development is promoted and a healthy, favourable and secure environment for economic activity and the region's residents assured. The strategy defines three major areas of activity in pursuit of the development vision and general goals: the Riga region must recognise its position as an area in Latvia and Europe, it must be competitive at the European level, with a diverse economy and an attractive business environment, and the quality of life must be enhanced. If the strategy is implemented successfully then by 2010 the Riga region will be transformed into:

- an influential Baltic Sea region, recognised throughout Europe and the world, where sub-regional authorities cooperate in pursuit of their local area development goals and regional goals;
- a region with a varied, innovative, and internationally competitive economy, offering an attractive business environment;
- a region with an expanding job market, rising income levels and improved welfare;
- a region in which residents enjoy a pleasant and secure living environment (Riga Region Development Council 1999).

The development strategy is the basic document on the adoption of consensus-based decisions and outlines necessary actions. A prerequisite for the long-term economic and administrative development of the region is the implementation of this strategy (Riga Region Development Council 1999).

References

Aizsilnieks A (1968) Latvijas saimniecibas vesture 1914-1945 (History of Latvian National Economy 1914-1945), Sundbyberg, Daugava

Boruks A, Kruzmetra M, Rivža B, Rivža P, Stokmane I (2000) Dabas un socialekonomiskie apstaklu mijiedarbiba un ietekme uz Latvijas lauku attistibu. (Interactions of environmental and socioecomical conditions and it impact on rural development in Latvia), Jelgava, LLU

Cabinet of Ministers Republic of Latvia (2000) Latvia's National Programme for Integration into the European Union 2000, Riga

CSBL – Central Statistical Bureau of Latvia (1996) Dzives apstakli Latvija. Norbalt dzives apstaklu petijums. Oslands O (ed) (Living conditions in Latvia), Riga

CSBL – Central Statistical Bureau of Latvia (1999a) Latvijas rajoni un lielakas pilsetas (Administrative Districts and Major Cities of Latvia), Riga

CSBL – Central Statistical Bureau of Latvia (2001) Living conditions in Latvia. Norbalt II project. An analytical report, Riga

CSBL – Central Statistical Bureau of Latvia (2000a) Macroeconomic Portrait of Latvia Regions, Riga

CSBL – Central Statistical Bureau of Latvia (1998) Socialie procesi Latvija (Social processes in Latvia), Riga

CSBL – Central Statistical Bureau of Latvia (1999b) Statistical Yearbook of Latvia 1999, Riga

CSBL – Central Statistical Bureau of Latvia (2000b) Statistical Yearbook of Latvia 2000, Riga

Dreifelds J (1996) Latvia in transition, Cambridge University press

Eglite P (2000a) Impact of Ageing on Human Capital in Latvia. In: Nordic-Baltic Sea region on the Eve of 21st century, Riga, pp 310-314

Eglite P (2000b) The Dynamics of Latvia's Urban and Rural Population at the End of the 20th Century. In: Folia Geographica Nr.7 pp 36-46

Eglite P, Markausa I (1993) Apvidu izvele paatrinatas attistibas veicinasanai (Regional policy in Latvia: selection of priority development areas). In: Latvijas Zinatņu Akademijas Vestis (Proceedings of the Latvian Academy of Sciences). Section A. Nr.12, pp 15-21

Francis I (2003) Transformation of Riga's territorial structure: processes, tendencies and results. PhD thesis. University of Latvia, Riga

Jankevics J (1991) Latvijas ekonomiska geografija (Latvia's Economic Geography), Riga, Zvaigzne

Jankevics J, Purins V, Jaunputnins J (eds) (1975) Latvijas PSR geografija (Geography of Latvian SSR), Riga, Zinatne

Kage I (2000) Transformacijas procesu izpausmes Latvija (Transformation Processes in Latvia), Riga, University of Latvia (unpublished)

Karnite R (1996) Pasvaldibu finansu izlidzinasanas metodika un budzeta aprekini (Methodology and Budget Calculations for Local Government Finance equalisation) Riga, Institute of Economics

Karnite R (1999) Reginal policy in Latvia. In: EU Enlargement, Cohesion and Regional policy. Strathclyde, EPRC University of Strathclyde, pp 131-142

Keiss S, Kazinovskis A (2001) Regionala administrativi teritoriala iedalijuma reformas problemas Latvija (Problems of the regional administrative division reform in Latvia), Riga, Local Government Department

Krišjane Z (1998) Dzives kvalitates teritorialo atskiribu izpete Latvija (Territorial differences of the quality of life in Latvia). Summary of PhD thesis. Riga, University of Latvia

Krišjane Z, Jankevics J (2000) Mobility of population and changing role of cities in Latvia. In: Nordic-Baltic Sea region on the Eve of 21st century. Riga, pp 125-130

Kruzmetra M (2000) Human Resources for Sustainable Rural Development in Latvia. In: Nordic-Baltic Sea region on the Eve of 21st century. Riga, pp 116-119

Latgale Regional Development Agency (2000) Latgale Urban Development Strategy, Daugavpils/Rezekne

Latvian Development Agency (2000) Investment in Latvia. In: www.latinst.lv/invest_latvia.htm

Liepaja district Council – Liepājas rajona padome, Liepaja city Council – Liepājas pilsētas dome, LR Vides aizsardzibas un regionalas attistibas ministija – Ministry of Environmental Protection and Regional Development Republic of Latvia, IRIS consulting, Pro-Inter CV (2000) Liepajas regiona telpiskais strukturplans (Spatial structure plan of Liepaja region) Izstrades metodika (Development method), Liepaja

Luse M, Grave Z (1993) Tehnisko un socialo infrastrukturu loma teritoriju attistiba (Kadu iesakņojusos uzskatu apstridot). (Factors of regional development in rural Latvia: closeness to urban centres vs. availability of local social service). In: Latvijas Zinatņu Akademijas Vestis (Proceedings of the Latvian Academy of Sciences). Section A. Nr.12, pp 6-14

Melluma A (2000) Latvijas pierobeža (Latvia's border area). Riga, Zinatne

Melluma A (1996) Materiali Latvijas reģionalas attistibas planam (Materials for regional development plan. Atlas) Atlants. Riga, Ministry of Environmental Protection and Regional Development, Republic of Latvia

Migracija (1998) Latvijas vesture. Datorenciklopedija (Migration. In: History of Latvia. Encyclopaedia CD), Riga

Ministry of Economy Republic of Latvia (1999) Economic Development Report of Latvia, June 1999, Riga

Ministry of Economy Republic of Latvia (2000a) Economic Development Report of Latvia, June 2000, Riga

Ministry of Economy Republic of Latvia – Latvijas Republikas Ekonomikas ministrija (2000b). Latvijas ilgtermiņa ekonomiska strateģija (Latvia's long-term economic development strategy), Riga

Ministry of Economy Republic of Latvia – Latvijas Republikas Ekonomikas ministrija (2000c) Latvijas rūpniecibas strateģija (Strategy of Latvia's industry), Riga

MoEPRD – Ministry of Environmental Protection and Regional Development Republic of Latvia (1998a). Latvijas regionalas attistibas politikas koncepcija (Regional Policy Development Concept). In: Regionala attistiba, Riga

MoEPRD – Ministry of Environmental Protection and Regional Development Republic of Latvia, (1998b) Valsts administrativi reģionala iedalijuma projekts (A Project Concerning the State's Administrative Regional Division), Riga

MoEPRD – Ministry of Environmental Protection and Regional Development Republic of Latvia (1999) Tūrisma nozares attistiba. Nozares parskats rajona planojuma izstradasanai (Development of tourism. Overview on elaboration tourism in district planning), Riga

Nikodemus O, Rozite M (2000) Tourism development in the context of sustainable development. In: Folia Geographica, Nr. 7, pp 69-76

Pocs J (1999) Regional, administrative and territorial reform in Latvia and conditions for its implementation. In: Institute of Economics (1999) Activities in 1999, pp113-124

PRO-INTER CV (Belgium), AGROTEC SpA (Italy), WES (Belgium) (2000) Pilot Regional Development Plan for the Latgale Region (May 2000) Draft final strategy. (EU Phare Contract: LE 9702.01.05/0001)

Republic of Latvia/European Commission (1999) Joint Assessment of the Economic Policy Priorities of Latvia, Riga

Republic of Latvia (2000) Strategy for the Integration into the European Union (approved by Saeima), Riga

Riga Region Development Council – Rigas regiona attistibas padome (1999) Rigas regiona attistibas strategija laika posmam lidz 2010. gadam. Projekts (Strategy for Development of Riga Region through 2010. Draft)

Rivža B, Rivža P, Kruzmetra M, Ramute L (1999) Valsts administrativi regionalais iedalijums: problemas, metodes, risinajumi (Administrative regional divisions of country: problems, methods and solutions), Jelgava

Rivža B, Rivža P, Ramute L, Kruzmetra M (1999) An assessment of administrative-regional reform in Latvia. In: Proceedings of the Latvia University of Agriculture. Volume 1 (295), pp 2-7

Rupnieciba (1984) Latvijas padomju enciklopedija. Vol 5/2, Galvena Enciklopediju redakcija (Industry, In: Encyclopaedia of Soviet Latvia). Riga, pp 341-370

Skinkis P (2000) Latvijas regionalas attistibas tendences un strategijas (Tendencies and strategies of Latvia's regional development). In: 2nd World Congress of Latvian Scientists. Congress Proceedings. Riga, Latvian Academy of Sciences, p 159

Skinkis P, Steins V (1992) Project of territorial division reform, Riga, Commission of the Administrative territorial reform. Supreme Council of Latvia (unpublished)

Spica I (2000) The Business Climate in Latvia. In: Nordic-Baltic Sea region on the Eve of 21st century, Riga, pp 108-110

UNDP (1998) Latvia Human development report 1998, Riga

UNDP (1997) Latvia Human development report 1997, Riga

Vaidere I (1999) Regionala politika Latvija: iespejas radit jaunas darba vietas (Regional policy in Latvia: opportunities to create new working places), Riga

Vaidere I, Vanags E (1997) The Promotion of Regional Economic Development. In: Humanities and Social Sciences, Nr. 2 (Latvia: State and Local Government as Promoters of Economy), pp 47-72

Vanags E, Vilka I (2000) Territorial Statistics and Regional Development in Latvia. In: Humanities and Social Sciences, Nr. 1 (Latvia: Latvian Economy at the Turn of Centuries), pp 17-34

Poland: Redevelopment Strategies at Different Levels

Elżbieta Okraska

Institute of Urban Development[1], Crakow

1 Development of Industrialised Cities and Regions in Poland

1.1 Historical Background of Industrial Development

1.1.1 History until 1918

Poland did not exist as a separate and independent state during the first period of European industrialisation in the late 18th and early 19th century. Its territory was divided between the three neighbouring states of Russia, Germany and Austria, and the process of industrialisation developed in different ways and with different speeds in these areas. In the middle of the 19th century, basic social and economic reforms were introduced on the divided polish territories to sweep away the remains of a feudal system that impeded industrialisation. The reforms included the following:

- enfranchisement of peasants (creating a cheap labour force and extending the domestic sale markets);
- technological revolution, i.e. shifting industries from the workshop stage of production to the stage of mechanised factory production (new technologies, new sources of energy and new raw materials);
- development of a railway network (leading to a reduction in transportation costs, with new supply and sales markets).

Relatively favourable conditions for the development of industry existed within the borders of the German state, due to the great successes of German foreign policy and rapid growth in the state-owned economy in the late 19th and early 20th century. The Polish territories annexed by Germany had large and diverse deposits of raw materials, in particular the area known as the Upper Silesia Coal Mine Basin (one of the world's largest coal mine fields), with a high level of capital concentration in the Upper Silesia Industrial Region. Also the Sudeten Industrial Re-

[1] Former Institute of Physical Planning and Municipal Economy, Crakow Branch.

gion, located in the Sudeten Mountains and its foot-hill areas, was at the time the most highly industrialised region in Europe.

The territories annexed by Russia had smaller, less varied deposits of raw materials and were consequently less industrialised. The main stimulants of industrialisation were the growing domestic sales market, the extremely absorbent sales market of the Russian empire and the strict import policy of the imperial government, protecting Polish and Russian industry against competitive western products by imposing import duty. In the 1880s, the textile industry centred in Łódź (the Łódź Industrial Region) became an important element of Russian industry in the Russian-controlled territory of divided Poland, with 43 % of all Polish textile workers concentrated in this area. At that time the city of Łódź was called the "Manchester of Continental Europe".

The slowest pace of industrialisation was seen in the territories annexed by Austria. One of the main obstacles was the strong competition from the highly developed Austrian and Czech industries, supported by the Austro-Hungarian monarchy. The Polish territories in the Austrian sector were among the least industrialised in the Empire.

It should be pointed out that the rate of industrial development in the annexed Polish territories was, in each case, much slower than that in the home provinces of the annexing countries. The structure of Polish industry before World War I was dominated by consumer-goods branches (61.2 % of the total industrial workforce), with the main components being the food industry (34.8 %) and textiles (21 %). The domination of consumer-goods industries is a characteristic feature of regions in their initial phase of industrialisation.

1.1.2 Inter-war Period

After World War I, the western part and some northern areas of modern Poland remained within the borders of Germany. The "Western Lands" were in a favourable economic condition, as industry there had not been destroyed during the war. The primary factors which impeded industrial development in the remaining territories (especially those annexed by Russia and Austria) were war damage, the loss of supply and sales markets (in particular the Russian one), a shortage of own capital, the war with the Soviet Union and a lack of economic and political stability after regaining independence. Generally speaking, the inter-war years saw very weak development of industry, with frequent economic fluctuations, weak conditions for growth and a global economic crisis in 1929. Existing industry was characterised by a domination of raw-material branches and outdated light industries (textile and food industries). Branches with the power to transform the national economic structure, i.e. metallurgy, electrical, machine and chemical industries, were particularly late in developing.

At the beginning of 1939, industrial centres were mainly grouped in the more industrialised south-western and central parts of Poland. Only a dozen centres were situated in the poorly industrialised regions east of the Vistula River, mainly created in the 1930s under the Central Industrial Region (COP) scheme. The formation of this new region began in 1936, following government initiatives in non-industrialised areas situated in the fork of the Vistula and the San rivers; several defence companies were located there for strategic reasons. The government also financed the development of defence, machine and motor industries in traditional old industrial centres, primarily in the Staropolski Region, Upper Silesia and Warsaw (Warszawa). After Poland regained access to the Baltic Sea, one of the greatest achievements of this period was the construction of a large port and industrial centre in Gdynia.

1.1.3 Socialist Industrialisation

At the end of World War II the borders of modern Poland were established. A new stage in the process of industrialisation began, differing fundamentally from that occurring in capitalist countries. The differences resulted from the state's particular social and economic systems, and from the fact that the basic means of production were nationalised. The Nationalisation Law of 1946 was aimed at industrial enterprises with more than 50 workers, while still preserving some private sector activity.

WW II had an even greater impact on the Polish economy than WW I. Modern industries located in Warsaw, the Western Lands and other centres of electrical, machine and chemical production suffered huge war damages; less hit were the textile, raw-material and coal industries. The development of industry in the People's Republic of Poland can be divided into several periods. The first, from 1944 to 1946, was a period of protection (preservation) and starting as well as preliminary reconstruction of industry, and saw the introduction of the Nationalisation Law and obligatory state management in vital enterprises.

Long-term economic plans were realised in the years from 1947 to 1975. The *Three-Year Plan (1947-1949)* was directed towards economic reconstruction, determining the partial extension of existing enterprises, primarily in the fuel and power industries (coal mines and power plants), metallurgy and motor industry; this was followed by the *Six-Year Plan (1950-1956)*, aiming for accelerated industrialisation, with priority granted to heavy industry, the wider distribution of industry, and the development of non-industrialised areas. Special emphasis was placed on the production of machines and equipment, electricity and raw-material extraction. Despite achieving an impressive industrial production growth rate of between 17.5 % and 27.7 % in the years 1950-1953, the plan was not fully realised; some of its assumptions were unrealistic and construction costs in non-industrialised areas were much higher than projected. Other plans followed, aiming in the late 1950s at the equal distribution of industry (a certain shift of investment projects from the southern macro-region to the central and peripheral ones was indeed obtained), and an increase of capital investment expenditures in industry, especially in the fuel and power generation industries. In the 1960s the aim was the continuation of accelerated growth in heavy industry. Developing branches were the bringers of technological progress, and in this period shipbuilding and machine building were developed on a large scale. Poland became a successful manufacturer and exporter of certain "turn-key" plants, e.g. sugar, cement, or sulphuric acid plants. At the time industry provided half of Poland's domestic product. However, the strenuous pace of investment and industrialisation adversely affected the dynamics of wage levels and consumption, i.e. the natural balance between the supply of consumer goods and the population's income was disturbed.

The characteristic feature of the 1970s was the broad international co-operation with both communist and capitalist companies. Poland was granted significant loans for the development and modernisation of various industries. Closer integration into the world economy stimulated the growth of export industries and selective industrial development. Changes were also seen in the structure of investment expenditures, with the share of the fuel, power generation and chemical industries falling, and the share of electrical, machine building, steel and food industries rising. At the end of the 1970s industrialisation in Poland was quite advanced and the industrial production rate was a little higher than the world average. However, symptoms of an imminent economic crisis could be detected from 1976, with particular problems arising in large projects such as the construction of the "Katowice" Steel Mill and wide-gauge "Steel and Sulphur" railway line from the Soviet border to the Upper-Silesia Industrial Region (GOP).

The 1980s were years of worsening economic recession and falling production in the majority of Polish industrial centres and regions. Despite some limited attempts at reform to the command economy, the absence of market mechanisms and poor links to the world economy resulted in a widening of the gulf separating Poland from the market-economy countries and also (in respect of numerous indices) from several of the countries in the former communist bloc. The unchanged basic economic mechanisms, in combination with the tense political situation of the 1980s, culminated in economic crisis in 1989.

1.2 Current and Future Development

1.2.1 Industrial Regions

The development of industrialisation in Poland can be divided into three general stages:

- systematic growth in the importance of the capital goods industry, with decreasing emphasis on the production of consumer goods;
- the concentration of industrial production in ever larger enterprises;
- the spatial concentration of enterprises in larger urban and industrial agglomerations.

The most conspicuous of these was the spatial concentration of industry, with industrial agglomerations (centres and regions) rapidly forming across the country. This resulted in the creation of 24 industrial regions, 12 of them arising before World War II (GUS 1994).

Different factors were involved in the creation of industrial agglomerations. Stanisław Misztal (1970) distinguished six regional types in Poland (new regions marked in italics):

1. regions with raw materials, either local materials or located in the direct vicinity, e.g. Górnośląski, Częstochowski, Opolski, Staropolski, *Karpacki*, *Piotrkowsko-Bełchatowski* and *Legnicko-Głogowski*;
2. regions with traditional home and workshop manufacturing systems existing before the appearance of factory industry: Łódzki, Bielski and Białostocki;
3. regions with both raw materials and a strong home and workshop manufacturing tradition before the introduction of factory methods: Sudecki and Zielonogórski;
4. regions mainly established in large cities, based on urban labour force, sales markets and infrastructure: Warszawski, Wrocławski, Krakowski, *Poznański*, as well as *Bydgosko-Toruński*, *Lubelski* and *Ostrowsko-Kaliski*;
5. regions established at large transportation nodes, based on the needs of the maritime economy: *Gdański* and Szczeciński;
6. regions related to a specific location: *Płocko-Włocławski*, *Rzeszowsko-Tarnowski* and *Sandomiersko-Tarnobrzeski* (the latter being associated with the extraction of sulphur).

The types of industrial regions as classified above show essential differences in the distribution and numbers of centres. Types 2, 4 and 5 are generally distinguished by the domination of one industrial region, often concentrating 50-90 % of total industrial employment. Thus, these are regions with a mono-centric industrial structure. In contrast, a characteristic feature of regions of type 1 or 3 (wholly or partly based on raw-material extraction) is a higher number of centres, with none dominating the others. Such regions are classified as having a poly-centric industrial structure.

The table shows the 32 largest industrial centres of Poland in 1985 which are also shown on the map. The largest accumulation of industrial centres is found in and around the coal-mining district of Upper Silesia. Five industrial centres – Górnośląski, Bielski, Częstochowski, Krakowski and Opolski – were merged into the single Górnośląsko-Krakowski Industrial Macro-Region. Covering only 6 % of Poland, but with 19 % of the country's population, this macro-region produces around 30 % of industrial output and has a corresponding share of the industrial workforce (Atlas 1996, Misztal); 20 large centres (from 220 industrial centres in the macro-region as a whole) each employ more than 20,000 people. The largest Polish urban and industrial complex was established there, with 45 coal mines, 3 zinc and lead mines, 16 steel works and 11 coking plants. Power stations, metallurgical plants, machine works, as well as the transportation, electrical and chemical industries are present, with centres in Katowice, Zabrze, Bytom, Gliwice, Chorzów, Ruda Śląska, Sosnowiec and Tarnowskie Góry. In majority of centres the fuel and power-generation industries and metallurgy dominate; machine and metal industries are the main branches in the remaining regions.

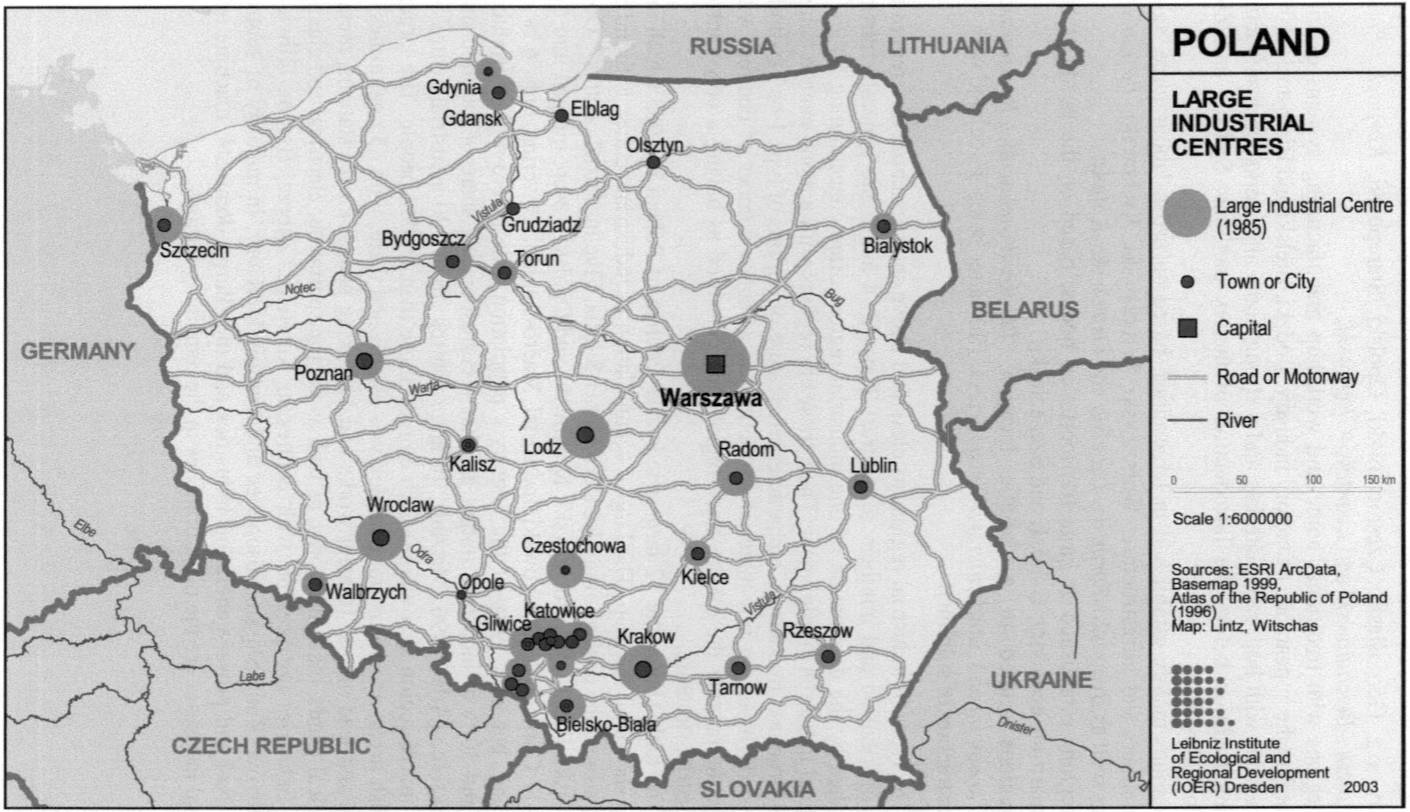

Map: Poland – The largest industrial centres 1985

Table: The largest industrial centres 1985, workforce in thousands

Centre		Centre	
Warszawa	234.4	Dąbrowa Górnicza	45.2
Łódź	170.5	Lublin	45.2
Kraków	118.5	Jastrzębie-Zdrój	44.5
Wrocław	100.4	Tychy	42.9
Poznań	89.0	Ruda Śląska	41.3
Katowice	88.4	Białystok	39.3
Gdańsk	71.5	Gdynia	38.0
Bydgoszcz	70.6	Toruń	37.4
Szczecin	59.9	Rzeszów	34.5
Gliwice	58.6	Kielce	34.2
Bytom	56.9	Rybnik	33.1
Bielsko-Biała	56.2	Tarnów	32.9
Częstochowa	54.0	Chorzów	31.7
Radom	52.0	Wałbrzych	30.4
Sosnowiec	50.6	Wodzisław Śląski	29.8
Zabrze	50.5	Kalisz	27.5

Source: Atlas 1996, City administrative borders in 1985

The second largest concentration of industry in Poland is found in the Warszawski Industrial Region. Apart from Warsaw, this region includes 60 smaller centres where industrial employment does not exceed 10,000 and the machine and metal industries dominate (with a considerable share of high-tech industry). Further agglomerations are the Łódzki Industrial Region, with a large share of light industry (mainly textiles), and the Sudeten Industrial Region, comprising about 130 scattered small and medium-sized centres with a considerable share of technologically outdated industries such as textiles, machine building, metallurgy and mineral extraction. A similar volume of industrial employment is found in the Staropolski Industrial Region, covering about 60 centres and including three larger cities – Radom, Kielce and Ostrowiec Świętokrzyski. In the majority of regional centres the machine-building and metal industries dominate. In 1980 the number of industrial centres in which the number of those employed in industry exceeded 1,000 workers was 590, and such centres concentrated 92 % of the industrial workforce.

1.2.2 Period of Transformation

The Economy

The collapse of the communist system in the Eastern Bloc countries in 1989 initiated a period of economic transformation which has continued for over 10 years. At the beginning of this process Poland had the weakest position of all. Throughout the 1980s, the gross national product per capita was lower than that of 1978. The period of ongoing transformation has included three stages (Stryjakiewicz 1999):

1. 1990-1991: initial transformation, "shock economy", "system breakthrough", a continued period of recession;

2. 1992-1993: overcoming recession and realising initial economic growth;
3. Since 1994: growth of macro-economic stability and industry.

In 1990 the institutional basis for transforming a highly-centralised economy to a market economy was established by the State-Enterprise Privatisation Act. Changes in the volume and structure of industrial production were primarily a result of the following stabilisation measures:

- transferring price control from the state to the market, leading to a price increase index of 1,154 in 1989-91,
- introduction of a restrictive remuneration and incomes policy; as a result real wages fell by around 25 % in 1990-91,
- internal convertibility of the Polish zloty, removal of state controls on foreign trade and lifting of import barriers, thus encouraging foreign competition that was difficult for domestic companies to counter,
- considerable reduction of subsidies (in certain branches these had previously exceeded the value of the manufactured products).

As already mentioned, the first two years of transformation was a period of continued recession; industrial production was affected by low domestic demand and high competition from imports. A rapid drop in output was reflected in a falling GDP – it fell 11.6 % in 1990 and by a further 9 % in 1991 (Atlas 1996, Zgliński). The situation could partly be blamed on the dissolving markets of the former Council for Mutual Economic Co-operation (COMECON), the dissolution of the Warsaw Pact and the associated demilitarisation of Poland's economy. At the beginning of the transformation period, 80 industrial enterprises were defence-related, providing around 200,000 jobs; by 1994 only 31 companies remained, with a total workforce of 80,000 (Stryjakiewicz 1999).

The regional adaptation to this stage of development was in two directions: decreasing employment in the industrial sector was accompanied by a relative increase in the service sector (in western, northern and central Poland), and farming underwent expansion (eastern and central Poland). Since 1992 the country has been out of recession: the value of production sold increased by 3.9 % in 1992, 6.4 % in 1993, 12 % in 1994 and 9.4 % in 1995, with GDP exceeding the 1989 level in 1995 (GUS 1999). Regarding branch structure, the beginning of the transformation period saw almost 10 % growth in output (compared to 1989) in the fuel and power generation industries, owing to a major increase in the price of power generation materials as well as the protective policy of the state.

The importance of food processing, wood, paper and mineral industries increased, while output of metallurgy, electro-machine and light industries decreased. These trends were halted during the second period of transformation. After the initial period of contraction, the electro-machine industry began to expand again, as did the wood, paper and food processing industries; this was reflected in the structure of employment.

Employment

The transformation of the state-owned economy led to fundamental changes in the volume and structure of employment. Within a short period of time the labour market in Poland changed from one characterised by labour scarcity (especially in industrial regions) to a market with low-labour demand. The number of jobs in the national economy (especially in non-agricultural sectors) fell dramatically, from 17.4 million in 1989 to 14.9 million in 1994. Northern regions (voivodeships) were particularly hard hit: the liquidation of state-owned farms contributed to a large rise in unemployment and brought a number of related problems.

Considerable changes also occurred in the structure of industry. The largest number of redundancies was in coal mining, machine, metal and light industries, with expansion only in the food-processing, wood and paper industries. As a result of these transformations, the private sector now had a dominating share of the workforce in the majority of regions. The least advanced areas in the privatisation processes were the Katowickie voivodeship (Upper Silesia Industrial Region, where 58 % of workers were employed in the public sector), Wałbrzyskie, Legnickie and Jeleniogórskie voivodeships (Sudeten Industrial Region). A considerably large growth in jobs was seen in the north-western part of the country, and in the Warszawskie, Łódzkie, Gdańskie and Elbląskie voivodeships.

Privatisation and Capital Investment

In 1990 the public sector provided 69.1 % of GDP, including 81.8 % of the value-added output in industry; by comparison, state enterprises in most highly-developed countries produce around 9 % of the average value-added output. The private sector in Poland has developed by the expansion of existing business entities and the establishment of new ones, as well as by the privatisation of state-owned enterprises. From 1991 to 1995 the number of industrial enterprises involved in the privatisation scheme grew from 221 at the beginning of the period to 1,821 at the end, constituting 35 % of state-owned enterprises. The total number of enterprises privatised between 1990 and 1998 came to 6,159. Ownership transformations and liquidation processes meant that the total number of state-owned enterprises decreased by 32 % in the period 1989-1994, from around 7,300 to 5,000.

The highest regional share of state-owned enterprises was maintained by the traditionally highly-industrialised voivodeships of Katowickie (9.5 %), Warszawskie (7.7 %), Poznańskie (6 %), Gdańskie and Wrocławskie (5.8 % each), while the lowest share was found in the voivodeships of south-eastern and central Poland.

After a period of reduced capital investment in the state-owned economy during the first years of transformation, a slow general growth in investment-related activities has been noted since 1992, with growing investment in industry since 1993. Capital-investment expenditures in the period 1992-1998 (in fixed prices) remained constant. Given the high inflation rate, one can discern a real and drastic drop in capital expenditures in mining and extraction industries. Foreign investment is of course an important factor in stimulating economic growth, modernising industrial structure, improving product competitiveness and increasing export volume. By the end of the first quarter of 1996, foreigners had invested 7,886.6 million US dollars in Poland (direct investments). That figure only totals investments made by the 3,777 companies which each invested at least a million dollars.

In fact the total value of foreign investments considerably exceeded 8 billion dollars, ranking Poland second in Central and Eastern Europe for foreign investment (in first place was Hungary). Those investments went mainly to large cities and urban agglomerations, with the Warsaw area getting the lion's share. However, a gradual trend of spatial de-concentration in the distribution of foreign capital has since been observed. Some large individual projects have been located in smaller centres and towns, such as Kwidzyń, Olsztyn or Dębica. Small towns with active local governments, such as Brzesko, Niepołomice, or Wolbrom, have also been the objects of investor interest.

The transformation processes in Poland's economy have bequeathed a spatial structure characterised by strong regional differentiation. The development of new private companies and the flow of foreign capital are most intensive in large cities and the outskirts of main urban agglomerations. In the region of Wielkopolska, factors such as a diverse economic structure, an opulent infrastructure and efficient agriculture have facilitated adjustment to the new market economy. Likewise, western voivodeships, using their location close to international borders, have positively encountered structural changes. However, economic transformation has been delayed and is causing some negative social fallout in just those areas which used to have a strong economic position: the mining and metallurgy zone of Upper Silesia, the Industrial Region of Łódź and areas of the Sudeten Industrial Region, as well as the regions which were previously dominated by state-owned farms.

1.2.3 Conclusion and Future Prospects

The following conclusions can be drawn from an analysis of the last ten years of transformation (Stryjakiewicz 1999): Structural changes in Polish industry confirm general trends in the world economy, with economies of scale losing importance in favour of economies of scope (shown, for example, by the development of small- and medium-sized businesses); the transformation of large public-sector enterprises to enable them to compete with competitors from highly-developed countries is a slow process, especially in mining and heavy industry; the globalisation of Polish industry (i.e. the flow of capital, technology and know-how) has been distinctly unidirectional so far. Multinational companies have assumed control of industrial branches such as automobiles, electronics/electrical engineering industries, food, tobacco, wood and paper processing, chemicals industry (especially cosmetics and cleaning supplies), and investments in the financial service sector. Unfortunately, international companies are not involved in those branches which require the most complex restructuring or largest capital investments. Foreign capital, despite numerous investment incentives, avoids regions which have been particularly hit by economic crisis. One positive effect of globalisation is the arrival of new forms of management and marketing.

Foreign investment has a positive impact on production output and improves adaptability indices. The influence of new institutional factors regarding industry location (investment incentives in areas with high unemployment, special economic zones, EU aid programmes) is relatively small in Poland. Preferred locations are persistently those in large urban agglomerations. The regions in which industry adapts best are Warszawski and Poznański, while the least capacity for adaptation is found in the Sudeten and Tarnobrzeski Regions. The main factors ensuring speedy adaptation are a diversity of industrial structures, a good infrastructure/institutional environment and social preparedness (entrepreneurial culture). Of course, the shift from a highly centralised economy to a market economy has accelerated the process of industry relocation from old run-down areas and facilities. Spatial changes were the result of transfers of ownership, as well as technological and organisational changes in enterprises, and such things as extensive (often unnecessary) ecological protection zones and more expensive/extensive utilisation of land in downtown areas. These processes have worsened the difficulties concerning the future of post-industrial areas and facilities, turning the problem from a local issue exclusively concerning mining areas to a national problem. The issue of post-industrial areas is closely linked to environmental degradation and pollution, and a transformation of these areas will require a new functional and spatial arrangement, i.e. a new state of balance between an area and its surroundings (Gasidło 1998). One indispensable pre-condition for any successful transformation is a healthy economy (generating demand for land for new capital investment projects), a well-developed science and technology sector, and a high level of social awareness and approval.

A comparative analysis of the diverse processes occurring during the transformation period using various indices is complicated by changes in terminology, different methods of statistical reporting, diverse source materials and a short analysis period. The database for industry is extremely poor and problems have resulted from the change of business-activity classification in 1993, i.e. the replacement of GUS State Economy Classification (KGN) by the European Classification of Activity (EKD), in which the concept of "industry" has been abandoned, replaced by three sub-categories:

- Mining and Quarrying (C),
- Manufacturing (D),
- Electricity, Gas and Water Supply (E).

A new administrative division of Poland was instituted in the Act concerning the Territorial Division of Poland of July 24th 1998, reducing the number of voivodeships to 16; this also complicates analytical research. All the above-mentioned factors mean that "the true picture of change to branch and type structures in Polish industry in the period of transformation is obscure, preventing us from drawing far-reaching conclusions" (Stryjakiewicz 1999).

It is therefore difficult to make projections regarding the further development of industry and industrial regions. Looking back one can say that in the 1990s the foundations for industrial market transformation were established in Poland. The economy stakeholders have been adapting to competitive conditions, as shown by the growth in industrial output which, since 1992, has been faster than in other countries of Central and Eastern Europe (Stryjakiewicz 1999). Up to now, growth has been based on the utilisation of simple reserves; maintaining that growth will require a long-term strategy to define the roles of companies and regions in industrial transformation as well as demanding the creation of a new spatial arrangement (with a decrease in the importance of traditional industrial regions based on raw materials and a cheap labour force). The direction of those transformations is also affected by the issue of Poland's membership to the European Union, as well as the continued institutional and legal development of the country's economic system.

2 Specific Problems and Reasons for Decline

2.1 Economic Problems

The shift from central planning to a free market economy has been very difficult. At the fall of communism, industry and industrial plants were concentrated in certain regions and used outdated technologies to produce low quality products. Coal-mining and metal production were particularly hit by the economic changeover, and of course the loss of the eastern markets badly affected most industrial enterprises. Many companies found it difficult to compete with international corporations. Aspects of communist economic management:

- a preference for heavy industry (consuming large amounts of energy and raw materials), reduction of consumer-goods industries and services, domination of the state-owned sector and of large enterprises (including industrial complexes) as well as a reduction in small and medium-sized businesses;
- arbitrary and non-economic grounds for deciding on business location, state-controlled determination of raw-material supplies and markets, low mobility of the means of production(especially labour, capital and technologies);
- system of prices and tariffs established by state authority, resulting in irrational transportation organisation and excessive use of vehicles, also lack of an independent financial sector and other institutions aiding business.

Communist Poland's share in the global market for labour and international trade was limited largely to one economic organisation – COMECON. The political transformations were accompanied by a dramatic reduction in the export of raw-materials (including coal and steel) and a collapse in shipbuilding and the defence industry. The main contractors had previously been within the Soviet Union and other Communist countries. The dissolution of the COMECON and the collapse of the USSR thus led to a huge loss of markets.

Post-communist problems regarding *branch structure* were the high degree of production and market monopolisation, the subsidising of the fuel and power generation industries and a simultaneous reduction of governmental expenditures on industry. From 1995-2000 the government reduced industrial expenditure by nearly 60 %. The amounts allocated to industry, calculated as a percentage of total budgetary expenditures, totalled 0.03 % (in 1995: 0.12 %).

One reason for decline on a *national level* is the weak and unstable legal system with frequent regulatory changes, often determined by the run-up to EU accession (particularly concerning taxation). One important example is 1994's Special Economic Zone Act, guaranteeing a fixed level of taxation benefits for 10 or 20 years. The Act had been amended by a 1997 Regulation establishing the priority of EU regulations over national law after Poland's entry into the EU (tax exemptions are not allowed under EU law; instead treated as a form of public subsidy to private business).

With the burden of its communist heritage, Poland has found the process of adapting its economy to the new rules more complicated and lengthy than ever projected. Negative factors have been:

- limited resources (in particular, financial), restricting a quick revival of capital and technology-consuming industries, and undermining international competitiveness;
- real advantages of Polish industry (cheap labour force) endangered by other countries;
- polarisation of Poland into economically stronger and weaker regions, with more industrialised western regions and less industrialised eastern lands.

The regions with economically diverse structures could adapt more easily than regions with mono-cultural economies or economies dominated by traditional branches. Thus the most advanced regions in the transformation process are easily distinguishable and associated with urban agglomerations (Warsaw, Poznań, Kraków, Wrocław, Bielsko-Biała, Gdańsk and Szczecin). These regions have a broad spectrum of economic structures, a population well prepared to face the free market, relatively low unemployment, high shares of the private sector, a well developed business environment, and the largest inflows of foreign capital. The one-time privileged regions, i.e. those with mining and processing centres dominated by traditional industry and raw-material branches, are slipping behind. The same is true of areas and cities whose economies were based on large defence-related enterprises, such as the Upper Silesia (Górnośląski), Sudeten and the Tarnobrzeski regions.

Areas with a history of weak and belated development, located in north-eastern Poland, show little sign of new economic vigour. The slowest development of the private sector is seen in these regions and little interest is shown by foreign capital investors. Insufficient local private capital, poor infrastructure and weak business environment institutions provide additional difficulties. An EU report shows that GDP per capita in the two richest voivodships in A-Poland (Śląskie, Mazowieckie) is currently ten times higher than in the two poorest regions in C-Poland (Lubelskie, Podkarpackie). In this categorisation A-Poland comprises the richest regions in north-western, central and southern Poland, B-Poland designates four regions on the western border and C-Poland comprises the five poorest regions on the country's eastern frontier.

2.2 Institutional and Social Problems

An unstable legislation and legal loopholes lead to insecurity and hesitancy on the part of business entities, as well as discouraging potential investors, fostering corruption and tending to expand the "grey zone" economy. Privatisation processes are clumsily handled and promising enterprises are being sold at less than fair market value. Proper market mechanisms are lacking and research and development is underfunded. Meanwhile the government is delivering weak macro-policies supposedly aimed at improving the market. All in all, it can be said that industrial regions do not receive the degree of interest they deserve and need from the Polish authorities. Frequent changes in government and a lack of agreement regarding even short-term political goals has resulted in a continual changeover in key personnel at all the levels, contributing to the incompetence of state officials and a growth in bureaucracy. Unemployment has led to internal and external migration and rising crime, and considerable resources need to be invested in education and training programmes. The widening gap in income between the poorest and the richest in the population has brought social unrest and increasing industrial unrest.

2.3 Environmental and Ecological Problems

The prolonged emphasis on state-run heavy industry has resulted in a badly polluted natural environment (water, air and soil). In the 1970s, the GOP area was one of the most highly polluted regions in Central Europe, while areas in the Upper Silesia and Kraków Macro-Regions, together with the Sudeten Industrial District, also suffered from high levels of environmental pollution. This legacy, compounded by the transformation processes, has produced approx. 6,000 km^2 (1996) of devastated or degraded waste-land requiring reclamation. Many polluted areas (such as abandoned production facilities) are to be found in city centres. The poor image of a given region and the essential deterioration in the quality of life are often the consequence of environmental pollution. On a positive note, some post-industrial areas located within town boundaries enjoy a more favourable position, benefiting from the restructuring or rehabilitation programmes.

3 Development Strategies

3.1 Objectives

The basic changes in the structure and orientation of social and economic development in Poland are the following: the reform of state administration to create the voivodeship as a regional policy entity; the arrival of a new stage in the integration process between Poland and the European Union; Poland's acceptance of the concept of sustainable development (recognized as an optimum strategy of economic development at the United Nations Conference in Rio de Janeiro in 1992). Sustainable development should allow the realisation of well-defined and socially desirable economic objectives, both long and short-term. Long-term objectives include

- rapid economic growth, with increasing GDP per capita;
- system stabilisation, favourable balance between supply and demand;
- ecologically sustainable development, maintenance of the natural ecological functioning of the environment, natural assets to be seen as long-term resources.

Short-term objectives are the following: improvements in legislation, with the elimination of inefficient and ineffective laws, increased transparency of industry activities, and greater sensitivity of social problems.

3.2 Strategies and Instruments

A *"Concept for Medium-Term Economic Development Until 2002"*, was approved by the Council of Ministers on June 15th 1999, having been drawn up by the Ministry of Economy in collaboration with scientific experts. This document combines a macro-economic vision of Poland's development with a prescription for the tasks and activities necessary to accomplish this vision. Specific tools and financial resources for implementation are defined in government programmes, some of which have already been approved. The document outlines the path to national economic growth for the next years, a particularly vital period in which social, economic and systemic reforms will be simultaneously implemented. These reforms will be decisive in preparing Poland for membership to the European Union. Economic policy established in the EU document aims to ensure compliance with the conditions for economic and monetary union. This is part of the pre-accession strategy under the "Partnership for Membership", and allows applicant countries to request all forms of EU assistance. The "Concept" is a starting point for the preparation of the National Development Plan up to 2006, and the elaboration of such a plan is a pre-condition for closer co-operation with the European Union in many fields, giving access to the Union's Structural Convergence Funds. The building of a modern economy, as indicated in the "Concept", is dependent on sustainable development, combining economic, social and ecological approaches. The goal is to create the necessary conditions to allow improved economic efficiency, while simultaneously preserving the cultural heritage and the natural environment. Increased innovation is one of the main factors contributing to economic growth, and the competitiveness of the Polish economy will in the future be determined by the ability to quickly transfer technological, managerial and organisational solutions to the business community, thereby ensuring commercial successes.

1999's *"Initial National Development Plan"* determined the main directions and priorities of Poland's social and economic cohesion up to 2002, with financing chiefly coming from EU aid resources. The document was approved by the

European Integration Committee on December 22nd 1999, allowing Poland to apply for PHARE fund resources under the social and economic cohesion programme. The goal of this programme is to support regional projects aimed at developing the manufacturing sector, infrastructure and human resources in selected regions and voivodeships. The necessary programme documents (Operating Programmes and Project Files) were prepared in 1999 by the Voivodeship Assembly Offices (local government administration) of the Śląskie, Podkarpackie, Lubelskie, Podlaskie and Warmińsko-Mazurskie Voivodeships, in close collaboration with the national government.

The *National Regional Development Strategy (NSRR) for 2001-2006* was approved by Resolution No. 105 of December 28th 2000. At the beginning of the 21st century, Poland is a rapidly developing European country, aiming for full social and economic integration into the community of advanced nations and, as an important step in this process, preparing for EU accession. At the same time, Poland still has a low GDP per capita in comparison to other European countries, and faces barriers to the development of a competitive economy and an improved standard of life for its residents. The situation is not uniform across the country, with particular regions and communities enjoying good development. The NSSR formulates a regional development strategy for 2001-2006, understood as part of the state social and economic development strategy, the intention of which is to realise the concrete steps necessary to attain objectives in national development, determined with respect to the territories of Poland and Europe. According to the Law Concerning Support for Regional Development, NSRR is one of the tools of the regional development policy system in Poland, determining objectives, priorities and criteria to identify areas needing support and to regulate regional support projects sponsored under the government's regional development policy.

This constitutes one point of reference when determining the allocation of state funds (supplemented by foreign resources) for regional contracts, for tasks pertaining to regional development, as well as consulting, information and pilot projects. The NSRR also specifies methods for coordinating actions by government ministers and specific funds in compliance with the objectives and tasks of the government's regional development policy, being directed for specific territories and included in regional contracts. Legislation establishes a basis for regional development programming at operating levels, and initiates specific measures and projects to stimulate regional development. NSRR is one of the basic medium-term structural strategies – other strategies include the national strategy for increasing employment and human resource development, the national strategies for developing rural areas and agriculture, the national fisheries strategy, the national environmental protection strategy and the national transportation development strategy. NSRR takes into account the provisions of the following documents, approved by the government to determine the objectives and directions of social and economic policy:

1. "Polska 2025" – A Long-Term Strategy for Sustainable Development (July 2000);

2. "Polska 2000-2010" – A Strategy for Public Finance and Economic Development (April 1999);
3. Concept for Medium-Term Economic Development in Poland until 2002 (June 1999);
4. Concept of Physical Planning Policy (October 1999);
5. Increasing Modernisation of Businesses in Poland up to 2006 (new technological, process, managerial and organisational solutions), focusing on the restructuring of mining, metallurgical and defence industry, as well as the restructuring of small and medium-sized businesses and domestic trade development;

The tools of NSRR strategy are: the stimulation of new capital projects, support for the development of small and medium-sized businesses, the establishment of regional investor support centres, the development of business infrastructure, direct financial support to businesses, the support of innovation generation and implementation (including modern technology transfers), the development of tourism and recreation, and the protection of the cultural heritage.

Poland suffers from a lack of structural funding, and therefore qualifies for inclusion in Objective 1 of the EU's structural policy. Community initiatives such as INTERREG and PHARE Cohesion Fund will be implemented. Legal documents relating to regional policy and development are the Support Programme and the Regional Contracts. The sources of NSRR financing include EU aid funds, Polish contributions to EU aid funds and an increase in the local-government share of public revenues and expenditures.

The *"Concept of National Spatial-Development Policy"*, prepared in 1999, was based on two earlier documents – the "Initial Concept" published in Report 3 in 1995 and the "Assumptions for a Social and Economic Development Strategy up to 2010", published in Report 4a in 1996.

According to the statutory formula contained in the Concept, the aim is to determine a nation-wide spatial policy. The term "spatial policy" is understood as a conscious shaping of the country's development to ensure particular strategic priorities. The Concept was prepared and updated under the aspect of a national development strategy. The provisions in Article 56 of the Land Management Act clearly specify that the Concept is a kind of study document, forming a non-binding spatial planning report with strategic significance. The assumption of such a formula is suitable for the needs of a free-market economy. In its current form the Concept is not a large-area spatial plan and does not determine the specific location of individual public capital investment projects. However, it is important to implement the Concept instruments. Specific programmes and government tasks serving to attain national public goals and affecting the physical development of the country are based on solutions established in the Concept. In this sense the Concept's role is to guide authorities when preparing the programmes under discussion. The remaining contents of the Concept are not binding and provide important data for other planning entities (Lexicon of Physical Economy 2000).

The basis for the planning system is the *Land Management Act* of July 7th 1994 (published in Dz. U. No. 89, Item 415). Amendments to this Act became effective

Łódź: The Green Ring of Culture and Tradition

Łódź is Poland's second largest city, located in the geographical centre of the country, 120 km south-west of Warsaw. The city was an important centre for textile production in the 19th century, being called the "Manchester of Poland" (Popławska 1986, 148). The industrial revolution brought an astonishing growth in the city's population – from 15,000 inhabitants in 1850 to 506,000 in 1913 – and liberal town-planning created an industrial coke-town with railroads, factories and slums. Post-war spatial development was marked by territorial expansion, central planning and application of the principles contained in the Charter of Athens; as a result the inner-city area degenerated. City development focused on the outskirts, resulting in a city core surrounded by a ring of modern apartments blocks and industrial districts. Inner city neglect has happily resulted in an almost perfect example of the European Commission's "idea of the city" (Commission of the European Communities 1990). Łódź can boast a multifunctional structure, and the degree of preservation of the 19th century architecture is still considerable. Political changes since 1989 have brought about the collapse of the local textile industry, formerly giving employment to over 50 % of the local labour force (Markowski and Kot 1993, 36). The number of inhabitants has been decreasing since 1990. Many production areas are contaminated and remain idle.

An "ecologically-beneficial urban development policy" was adopted as the main principle of the 1993 City Master Plan. The renewal of the urban core and enhancement of the city's identity were considered especially important for ensuring sustainable development in Łódź. The collapse of industry gave a rare possibility of preserving the city's industrial heritage and industrial identity, at the same time as obtaining good ecological development. Thus, the 1993 City Master Plan introduced the idea of the "Green Ring" of Culture and Tradition. This Ring incorporates the green areas of existing city parks, cemeteries and industrial areas situated in two valleys of the rivers Łódka and Jasień. The valleys were designated as "ecological zones", forming basic elements in the regional environmental structure (an idea comparable to the Green Corridors of the Emscher Park). The river valleys were also designated as "ventilating corridors". A recognition of the river valleys as protected areas which enrich the city's green belt (as determined in the so-called City Identity Area) is fundamental to the city's sustainable development (Łódź City Office – Department of Environment Protection 1997, 155). The authors of the 1993 City Plan received an award from the Minister for Town Planning and Construction for their work.

Unfortunately the city of Łódź has not been rigorous in following the plan. For example, in the northern part of the Green Ring a former industrial area was designated for conservation in 1971 (Szymański 1997), and subsequently given the status of a City Park in the 1993 City Master Plan. However in 1999 the city authorities altered the City Master Plan, allowing a French developer to buy the area for the construction of a hyper-mall.

Przemysław Szymański, Technical University Łódź

on January 1st 1999, determining the competencies of state administration agencies in relation to systemic state reform, especially where physical planning procedures are affected.

According to the changes introduced, the physical planning procedure is the following: On the national level, the "Concept of the National Spatial Development" is prepared together with periodic reports on the country's developmental conditions and programmes specifying the government's tasks to fulfil public objectives. On the regional level, voivodeship assemblies approve the following planning documents:

- "The Strategy of Spatial Development";
- "The Physical Plan of the Voivodeship";
- medium-term voivodeships programmes;
- foreign co-operation priorities of the voivodeship.

The voivodeships programmes should comply with the "Concept of the National Spatial Development" and the "Physical Plans" of neighbouring voivodeships. Neither the "Physical Plan of the Voivodeship," nor any of the other documents specified above are local ordinances, and they do not breach the rights of communes. At the county level (*powiat*), authorities may conduct non-binding analyses and studies in physical development within their substantive jurisdiction. Two documents are prepared at the commune level (*gmina*):

- "A Study of Conditions and Directions of Land Management in the Commune";
- "The Local Physical Plan" (for the whole or part of the commune).

A projection of the Plan's impact on the natural environment is included in the Local Physical Plan, prepared by an expert certified by the Ministry of Environmental Protection, Natural Resources and Forestry. Of the documents specified above, only the Local Physical Plan has a legal basis (at local level). An essential procedural element is the holding of negotiations between the voivodeship and communal authorities concerning the non-local state tasks in Local Physical Plans. The presented system requires modification, by either amending the existing Land Management Act or approving a new one. This task is even more urgent in view of the fact that on January 1st 2001 the Local Physical Plans approved before January 1st 1995 became invalid.

Apart from spatial policy, *sectoral policy programmes* were also determined. In 1995 the Ministry of Industry and Trade prepared several industrial-policy programmes to encourage the development of small and medium-sized businesses, including:

- "Analysis of the effectiveness of industrial and technical policies, with proposed modifications";
- "International competitiveness of Polish industry. An industrial policy programme for the period 1995-97";

- "Elements of state policy concerning the trading of goods and services, affecting Polish economic competitiveness, in the light of the European single-market principle";
- "Opportunity areas: identification and procedural concepts".

In June 1998, following some unsuccessful concepts (including those of April 30th 1996), implementation began of the "*Programme for Coal Mining Reform in Poland, 1998-2002*". The main goal of the reform was to achieve profitability in coal-mining corporations, including measures to reduce capacity and personnel, with the closure of selected mines. However, these actions can of course bring an increase in structural unemployment. This is a realistic threat to mining reforms, as well as to the maintenance and development of opportunities for the whole region. Therefore, the "Law concerning the adjustment of coal mining to market conditions" of November 26th 1998, with special entitlements for mining communities, provides organisational, legal and financial possibilities for projects intended to revive local economies, obliging mining enterprises and local communities to co-operate.

In addition to development strategies and sectoral policy programmes, legal steps to ensure institutional support were taken, both centrally and locally. These included an investment incentive which reduces income tax for investors who bring new jobs to communes with structural unemployment. The Act of October 20th 1994 established *Special Economic Zones (SSE)* in regions capable of accelerated development or those particularly struck by economic crisis, giving a boost to economic development (e.g. by making better use of existing industrial infrastructure). The Special Economic Zones are similar to those previously established in Western Europe. The Special Economic Zone Act guarantees taxation privileges for 10 or 20 years and strictly determines the type of business which may be developed in such zones. The zones are managed either by the Agency of Industrial Development or local companies, with shares owned by the State Treasury.

The restructuring efforts were supported by several EU initiatives. The PHARE STRUDER programme was the first regional programme in Central and Eastern Europe to help areas with outdated industrial and agricultural structures. In Poland this programme has been implemented in the industrial centres of the Upper Silesia (Górnośląski) (the former Katowickie voivodeship), Sudeten (the former Wałbrzyskie voivodeship) and Łódzki Regions. The EU project PL9811 – "Amelioration of Social and Regional Effects of Coal Mining and Metallurgy Restructuring" – was introduced by the Polish Foundation for Small and Medium-sized Enterprise Promotion and Development under the supervision of the Ministry of Economy. In addition, Poland has benefited from the 1999 PHARE Programme for Industry Modernisation, the Special Accession Programme for Agriculture and Rural Development (SAPARD), and the Instrument for Structural Policies for Pre-Accession (ISPA).

In 1999 the Special Programme of Preparation for Structural Funds (SPPSF-PHARE '98) was introduced. The chief objective of this Programme is to prepare central and regional government officers for the necessary programming and implementing of regional policy, both before and after Poland's accession to the

European Union. The preparation of regional development strategies has been supported by a specific reserve fund designated for the co-financing of the SPPSF. A strengthening of local administration has also been financed from resources granted by PHARE '97, Integrated Regional Development (INRED) and European Integration programmes. Bilateral co-operation included training, workshops and study tours to the member countries of the European Union.

Further foreign support for restructuring included the Polish-British Entrepreneurship Programme, set up in co-operation with the Know-How Fund and the World Bank. The Bank's team of experts, acting jointly with representatives of the BSW SA (Bytom Coal-Mining Company), prepared the "Programme for Partial Closure and Development of BSW SA Coal Mines".

When considering the recent strategy and transformation projects, it should be remembered that similar activities were implemented in Poland long before 1990, in particular in industrial areas affected by pollution. One of the earliest local projects of regional importance was the construction of a Regional Culture and Recreation Park in Chorzów in 1952. The Park was built in the very heart of the Katowice industrial centre, and showed the possibilities of transforming devastated and hazardous dumping sites or collapsed mine fields into popular facilities for active and passive recreation.

The best known regional programme involving the transformation of post-industrial areas was the development of the GOP (Górnośląski Okręg Przemysłowy – Upper Silesia Industry Region) Protective Forest Belt. The project, conducted since 1968 by various administrative and business entities, was managed and financed at regional and national level. The aims were to improve the environment in the direct vicinity of the Katowice industrial centre, with the enhancement of factors influencing climate and health, more effective forest management, the development of recreation areas for inhabitants of Silesian towns, and the reclamation of abandoned sand pits and other post-industrial areas located within the Belt. The last two objectives were realised by transforming sand pits into artificial lakes featuring weekend recreation facilities, e.g. in Czechowice, Pławniowice, Chechło and Pogoria.

Having analysed the main factors which determine the future prospects of post-industrial towns and regions undergoing structural transformation, we are now able to draw the following conclusion. Bottom-up activities – i.e. local projects pursued in communes, counties and regions, with contributions from both local and regional institutional partners (e.g. local governments, national agencies such as the Agency for Regional Development, Coal Corporations, the Agency for Local Initiatives, the Industrial Development Industry, or the Environmental Protection Agency) – can use the tools of national, regional and sectoral policies to prepare effective development programmes, with government support in legislation and financing (aid programmes). These programmes will be understandable and acceptable to all parties, and therefore should encounter no resistance to implementation.

References

Atlas of the Republic of Poland (1996) Atlas Rzeczypospolitej Polskiej, Warszawa, with articles of Misztal and Zgliński

Changes in the Enterprise Ownership Structure and in Industrial Production in 1993-1994 (1996) „Zmiany w strukturze własnościowej przedsiębiorstw i produkcji przemysłowej w latach 1993-1994", GUS, Warszawa

Concept of Medium-Term Economic Development of Poland Until 2002 (1999) „Koncepcja średniookresowego rozwoju gospodarczego kraju do roku 2002", Ministerstwo Gospodarki, Warszawa

Concept of Regional Development (1986) „Koncepcja rozwoju regionalnego", Seria Zarządzanie i Marketing, Częstochowa

Fierla I (1984) The Polish Industrial Economy. "Gospodarka przemysłu Polski", PWE Warszawa

Gasidło K (1998) Problems of Transformations of post-industrial Areas. "Problemy przekształceń terenów poprzemysłowych", Architektura z 37, Zeszyty Naukowe Politechniki Śląskiej, Gliwice

GUS – Central Statistical Office (1994) Delimitation of Industrial Regions – Territorial Changes in 1978-1988, "Delimitacja Okręgów Przemysłowych – zmiany terytorialne zaistniałe w latach 1978-1988", Warszawa

GUS – Central Statistical Office (1985) Delimitation of Industrial Regions Delimitacja Okręgów Przemysłowych,

GUS – Central Statistical Office (1999) Statistical Yearbook, Rocznik Statystyczny Warszawa

Industrial Regions in Poland (1985) Okręgi przemysłowe w Polsce, GUS, Warszawa

Kortus B (1986) Introduction to Industrial Geography. "Wstęp do geografii przemysłu", PWN, Warszawa

Lexicon of Physical Economy (2000) "Vademecum Gospodarki Przestrzennej", rękopis, IGPiK O/Kraków, Kraków

Lijewski T (1978) Industrialisation of Poland 1945-1975. Structural and Spatial Changes. "Uprzemysłowienie Polski 1945-1975. Przemiany strukturalne i przestrzenne", PWN, Warszawa

Misztal S (1970) Changes in the Physical Structure of Industry in Poland, 1960-1965. "Przemiany w strukturze przestrzennej przemysłu na ziemiach polskich w latach 1960-1965", tom XXXI, PWN, Warszawa

Model of Transformations of Post-Industrial and Degraded Areas (1999) "Modelowe przekształcenia terenów poprzemysłowych i zdegradowanych", redakcja Krzysztof Gasidło, Justyna Gorgoń, Katowice

Pakuła L (1997) The Upper Silesian Industrial Region (in German). ERDKUNDE Unterricht 12/96, 1/97, "Der Oberschlesische Industriebezirk (Teile I, II)"

Policy Concept of Country's Physical Development (1995) "Koncepcja polityki przestrzennego zagospodarowania kraju", Rządowe Centrum Studiów Strategicznych, Warszawa

Revitalisation, Rehabilitation and Restructuring/Renewal of Towns (2000) Rewitalizacja, Rehabilitacja, Restrukturyzacja/Odnowa miast – praca zbiorowa, IGPiK O/Kraków, Kraków

Strategy of Integrated Development of Katowice Voivodeship 1998-2002 (1997) "Strategia zintegrowanego rozwoju województwa katowickiego 1998-2002", Katowice

Strategy of Development of Małopolska Voivodeship for 2000-2006 (1999) "Strategia rozwoju Małopolski na lata 2000-2006", Kraków

Strategy of Economic Development of Kraków Voivodeship (1998). "Strategia rozwoju gospodarczego województwa krakowskiego", redakcja Jan Targalski, AE w Krakowie, Kraków

Stryjakiewicz T (1999) Spatial Adjustment of Industry in Poland under Transformation. "Adaptacja przestrzenna przemysłu w Polsce w warunkach transformacji", Wydawnictwo Naukowe UAM, Poznań

Wrzosek A (1972) Main Industrial regions of Poland. "Główne Okręgi Przemysłowe Polski", PZWSz, Warszawa

References of Case Study:

Commission of the European Communities (1990) The Green Paper on Urban Environment, COM (90)214, Brussels

Łódź City Office – Department of Environment Protection (1997) Ecological Policy for the City of Łódź-Local Agenda 21, Łódź

Markowski T, Kot J (1993) Planning for Strategic Economic Development of Łódź - concepts, problems and future vision of a city, Kronika m.Łodzi, Zeszyt 1/93, Łódź

Popławska I, Muthesius S (1986) Poland's Manchester: 19th-Century Industrial and Domestic Architecture in Łódź. Journal of the Society of Architectural Historians. June 1986, vol. XLV n. 2, pp 148ff

Szymański P (1997) Die Poznanski-Werke (in German). Stadt Bauwelt 136, issue 48, pp 2730-2735

Romania: Consequences of Small Steps Policy

Serban Nadejde, Dorottya Pantea, Adrian Dumitrache and Ionica Braulete

National Institute for Research and Development in Urban and Regional Planning, Bucharest

1 Situation of Industrialised Cities and Regions in Romania

1.1 Development in the Past

Industry in Romania has developed in logical steps, in accordance with the type and geographical distribution of natural resources. For example, mining activities have been recorded in certain areas since Roman times. In the Middle Ages, smelting, textile production and carpentry were very much small-scale industries. At the beginning of the 18th century, oil extraction and the use of water mills opened the way to factory production, and a truly "modern" type of industry was beginning to form at the end of the 18th century. Against a background of the growing importance of the extractive sector (oil, coal, ores), more traditional branches also saw considerable expansion (textiles, food, paper and building materials). In the meantime, branches such as industrial chemical production and modern wood processing were added. The emergence of large-scale industry was backed up by the building of a railway system, and also shaped by new forms of apprenticeship and training. In the inter-war period, the main industrialised areas of Romania were:

- Bucharest and its vicinity (complex activities, especially oil drilling and processing);
- Central and south-eastern Transylvania (energy production, metallurgy);
- Central Banat (machinery and equipment);
- Western Banat (various manufacturing branches);
- Western Maramures (mining and processing of non-ferrous metals);
- Crişana (complex activities);
- North-eastern Bărăgan (shipping); and
- Western Moldova (chemistry, pulp and paper industry).

The first coal mines began production in the late 18th century. By the end of the 19th century massive coal extraction areas such as the Jiu Valley and the Almas Basin were established. Coal output jumped from 64.8 tons in 1889 to 104,300 tons in 1900, and with modern rail transportation, further increased to 300,000 tons in 1915; however in 1938 production dropped to 3,000 tons. The industrial exploitation and processing of oil began in the middle of the 19th century. In a short time Romania was the main oil producer in the world: by 1900 production stood at 100,000 tons, rising to 1.5 million tons in 1910 and 1.59 million tons in 1915. The inter-war period saw great fluctuations in production: 8.7 million tons were produced in 1936, falling to 6.6 million in 1938 and 3 million in 1946. The petrochemical industry was mainly located in the Prahova Valley.

From the first decades of the 19th century impetus was given to engineering by smaller workshops manufacturing agricultural and mining equipment (later machinery for textile and food industry) in Banat, Transylvania and Moldova. In Bucharest, large factories with over 500 employees were established, sometimes using foreign capital. Shipyards and factories producing heavy metal equipment (e.g. for the railways) were built in the second half of the century. Before World War I 83 different enterprises were engaged in metal processing. In 1938, the number of such companies was 366, providing 10.2 % of the country's industrial output (third place after food and extraction industries). Chemical industries, utilising Romania's natural resources, first appeared at the beginning of the 20th century. The main products were sulphuric and nitric acid, ammonia and caustic soda. In the second half of the 19th century a modern pulp and paper industry emerged, with most factories located in mountainous areas, where forests provided raw material. The inter-war paper production was sufficient to meet internal demand and provided significant exports.

The Socialist era after World War II was characterized by a typical, centrally-planned economy, with the forced industrialisation of all regions of the country, and agriculture also being put under central control. Emphasis was placed on coal extraction, realising outputs of 38 million tons in 1980 and 66 million tons in 1989. Unfortunately the physical and social infrastructure to sustain this branch did not receive sufficient investment over the long term. Pre-war oil refineries were rebuilt, modernised and specialised for different oil-related products. In the 1960s, new refineries were constructed in the South and in the East of the Carpathians, and in 1979 the platform Midia-Năvodari began operations. After 1960, crude oil output increased, peaking in 1974 (14.7 million tons) and subsequently dropping to 9.2 million tons in 1989. The extraction of natural gas, exploited since 1908 in the Transylvania Plain, was extended to newly discovered deposits in the Romanian Plain and in the Getic Plain, providing the chemical industry with an important source of energy. The chemical and the petrochemical industries were supported by the state, production was diversified and many new plants were built. Unfortunately, some of these were located far from the site of resources and others relied on imported raw materials. Food and textile industries, providing goods both for the internal market and export, were present in almost every important urban centre.

In the socialist period, six economic macro-zones were determined (according to the Economic Geography of Romania; Cucu 1996):

- the Southern Peri-Carpathian area, comprising old and new industrial centres of extraction (oil, coal); petrochemical industry, iron production, drilling equipment and engineering plants, shipyards and chemicals;
- the South-Western Banat area, including centres of mining, metal processing and engineering;
- the Crişana-Maramureş area, developed on the basis of local energy, non-ferrous and wood resources; area contains several extractive, metal and wood processing centres, as well as engineering industries;
- the Eastern Peri-Carpathian area, where a "young" industry emerged, with energy generation plants, metallurgy and engineering, chemistry;
- the South-Eastern area, also with dominantly new industry, developed on the basis of imported raw materials and shipping. Here we find centres of metallurgy, chemistry and shipyards;
- the Central Transilvania area, with dominantly old industrial centres, based on regional energy and raw material resources and extensive manufacturing (e.g. engineering, chemistry, construction materials).

1.2 Present Situation

1.2.1 General Outline

In the 1990s, the evolution of the Romanian economy was marked by the transition to democracy and a free-market system, with prolonged economic decline, strong structural pressures, belated restructuring of the real economy and, consequently, a background of growing social tensions. The progress of economic, social and institutional reforms was hampered by the lack of coherence and coordination of its components, but also by the authorities' reduced ability to manage the crisis; poor communications between the social partners did not help. In the last 7 years, economic developments have been influenced by factors reflecting different, even contradictory, visions regarding the transition to a market economy. From 1993 until 1994, monetary instruments were mainly used to realise economic stability, without significant changes in the real sector being attempted. Between 1995 and 1996, electoral considerations checked any effort towards economic restructuring. Thus, against a temporary improvement in the quality of life, the country's external debt increased and the privatisation process slowed down. From 1997-2000, the new government initiated rough reform measures, but these succeeded only partly in mitigating the downward spiral. Economic decline became acute and the population's confidence in the authorities' ability to manage the situation diminished. The Balkan war, with the subsequent embargoes (until now Romania has not received any compensation), as well as a long-lasting drought, have exacerbated the problems of transition.

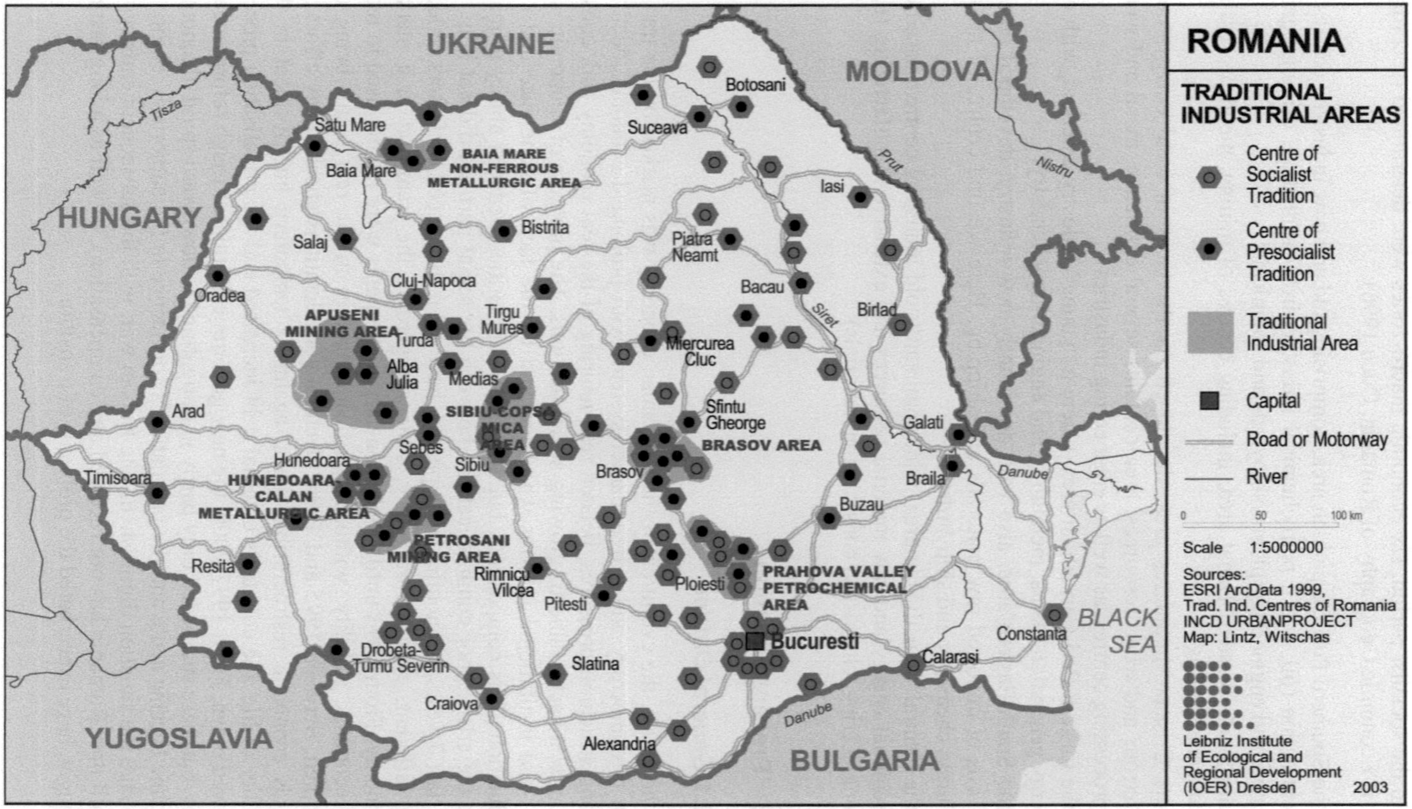

Map: Romania – Industrial Centres and Areas

1.2.2 Industrial Centres in Romania

The definition of an "industrial region" was a difficult task, because:

- data concerning the volume of production are confidential at the level of the economic unit and are not centralised at the level of the administrative-territorial unit;
- most private enterprises have a mixed profile (manufacturing and services), and their activity fluctuates in line with demand. That makes it difficult (using official data) to determine the real economic profile of an area;
- the relatively small supply of domestic capital, together with high tax levels, have encouraged the emergence of a grey economy which can only be estimated in size.

For the present study, the notion of "industrial centre" was defined by Urbanproiect as "an administrative-territorial unit where the share of industrial employment in regard to total employment is dominant". Additionally, the disadvantaged mining areas were considered according to the laws in force. Industrial centres are therefore grouped in "cities" (also comprising some communes) and "regions" (comprising disadvantaged areas and adjoining industrial centres). Among the localities with important industrial functions, 135 genuine industrial centres (with dominant industrial function) of two types were identified: The capital city and county residential municipalities, where industrial employment exceeds 35 % of total employment (Type A Centres) and municipalities, towns and communes with over 50 % of total employment being industrial (Type B Centres). Table 1 shows the different types of industrial centres in certain regions of Romania.

Table 1. Industrial centres in Romania

Regions of Development	Type A Centres	Type B Centres	Total
North-East Region	6	10	16
South-East Region	4	4	8
South Region	5	19	24
South-West Region	5	6	11
West Region	3	15	18
North-West Region	6	9	15
Centre Region	6	29	35
Bucharest-Ilfov Region	1	7	8
Total	36	99	135

Source: INCD Urbanproiect Bucharest

There are exceptions to the aforementioned rule, in the case of centres with industrial units of national importance, or the co-ordinating centres of some well-defined industrial areas, where industrial employment does not reach the quoted share. Data concerning employment refers to 1995 (1994, for Argeş county), which was the most recently available (source: Localities files, National Commission for Statistics). We estimate that in the period 1996-2000 the employment level dropped by 25 % in some areas. Among the 135 industrial centres, 75 have a longer pre-socialist tradition, grouped mainly in the centre and in the north-west of the country (see map).

The share of "socialist" industrial centres (44.5 %) is very high, which shows the determination of the communist regime to promote industry as the main economic branch. The territorial distribution is relatively balanced, with the exception of some plain (agrarian), mountain (low accessibility) and delta (protected) areas. Areas with a higher industrial density are:

1. Apuseni mining area;
2. non-ferrous metallurgy area Baia Mare;
3. chemicals area Sibiu-Copşa Mică;
4. petrochemical area Prahova Valley;
5. metallurgy area Hunedoara-Călan;
6. coal mining area Jiu Valley;
7. Braşov complex area.

1.2.3 Features of the Economy

Sectoral Structure of Economy

As can be seen in Table 2, national figures for industrial production show a steep downwards trend between 1990 and 1992, growth beginning in 1993, and sharply dropping again in 1997 as a consequence of the government's radical restructuring policies.

Table 2. Change of level of industrial production (%)

	1990	1991	1992	1993	1994	1995	1996	1997	1998
Change compared to previous year	-23.7	-22.8	-21.9	+1.3	+3.3	+9.4	+6.3	-9.8	-17.0

Sources: Ministerul Industriei şi Comerţului (1996), National Commission for Statistics 1998

From 1990 to 1999, the absolute figures for employment dropped by 16.8 %, (of which the decrease in industry was 39.0 %). Yet the growth of the number and share of workers in the primary sector is significant. This reflects the absorption of industrial unemployment by agriculture (Table 3). Employment in agriculture increased both in absolute terms and as a percentage of total employment.

Table 3. Structure of economy by employees (%)

	1992	1993	1994	1995	1996	1997	1998
Total	100.0	100.0	100.0	100.0	100.0	100.0	100.0
Primary sector	32.9	35.9	36.4	34.4	35.4	37.5	38.0
Secondary sector	37.1	35.8	34.4	33.6	34.3	32.0	30.7
– of which industry	31.6	30.1	28.8	28.6	29.2	30.7	26.3
Tertiary sector	30.0	28.3	29.2	32.0	30.3	30.5	31.3

Source: National Commission for Statistics, 1991-1999

Agriculture and forestry, having a share of about 18 % of GDP (1998), were the only dynamic sectors in the national economy. Against the background of economic structural adjustments and the linking of production to the market, agriculture acts as a buffer in meeting consumption demands and absorbing many of the people who had previously been employed in industry. Since 1993, agriculture and forestry have dominated the structure of employment in Romania (38.0 % in 1998, Table 3). There are 3.9 million private farms, each with an average area of 2.3 ha/farm. This fragmentation covers 84 % of the agricultural land in Romania and represents one of the major problems of the sector. The unconsolidated nature of farming is also illustrated by data concerning the structure of crops and the reduced efficiency in general. The large number of older people working in this sector also means that resistance to innovation is considerable. Subsistence farming is widespread.

Under the term "grey economy" the authors understand "unregistered and/or underevaluated economic activities" (Editura Expert 1998). The grey economy increased by a factor of 3.3 between 1992 and 1998, rising to over 21 % of GDP. Estimations regarding the percentage of GDP made up by the entire shadow economy (including other activities, some of them illegal – black economy) in 1998 range from about 40 % of GDP (estimation by the Romanian Information Services) to 50-60 % of GDP (estimation by specialists from the Ministry of Industry and Commerce).

Productivity, Foreign Investment and Privatisation

Compared with the starting point of transition, labour productivity increased, especially in 1994 and 1995, without compensating for the collapse of activities registered in the first years. The general trend of GDP per worker is positive (it almost doubled from 1990 to 1996), yet, the GDP per capita decreased again after a slow increase until 1996. This indicates that even if productivity in general has risen, GDP has dropped and the number of workers dwindled. The largest increase in productivity was registered in the high-tech industry (over 140 % from 1995 to 1999). At the same time, a constant loss of productivity was seen in the textile industry, chemical industry, metallurgy, machinery and equipment.

Foreign investments up to 1999 show a sluggish pace: at the end of the year they amounted to 4.94 billion USD (Dec. 1990-Dec. 1999), that is 194.1 USD per capita. Foreign investments are concentrated in four regions: Bucharest and surroundings, the north-western and central part of the country, Constanta County (major gateway to the Black Sea) and the Craiova region. Because of its particular economic structure and the delay in applying structural reforms, Romania is greatly dependant on external trade. In the last 10 years, imports have consistently exceeded exports; the import surplus increased from 63.9 % in 1990 to 80.9 % in 1997. The data concerning exports show, on one hand, a general decline, with a slight recovery seen in 1999. On the other hand, products with a low value added are increasingly finding a market abroad. As the share of total exports made up by industrial products consistently exceeds 92 %, one can conclude that in recent years Romania has been exporting mainly low-technology industrial goods.

Since 1990 private capital has been created through the privatisation of state enterprises and founding of new private companies. Privatisation of state-owned enterprises gave rather limited results up to 1995. Over this period 1482 companies were privatised, mostly small businesses (1089). From 1996 this privatisation accelerated, coming to an end by about 2000. The majority of economically active enterprises is now in private ownership, and the share of the private sector in terms of GDP has been on the increase since 1990, reaching 61 % in 1998.

Employment and Social Aspects

In recent years employment levels in Romania has been continually worsening: in 1992 the number of those in work was 10.8 million, representing 45.9 % of the population, while in 1998 the figure was 8.8 million (39.2 % of the population). The geographical distribution shows low levels of employment in the north-east, south and south-east regions, compared with a higher employment level in the central and north-western regions. Unemployment is comprised of two groups: the "registered unemployed", officially recognised, and the ILO unemployed. Both figures increased year by year up to 1998, showing stability or even a slight downward trend in 1999 (Table 4). Yet unemployment figures are relatively small in comparison with other former socialist countries. Rather than signalling a better economic situation, this can be attributed to the delayed introduction of economic reforms and the maintenance of many loss-making industrial units.

Table 4. Changing unemployment

	1992	1996	1997	1998	1999
Registered unemployment rate	8.2	6.6	8.9	10.4	10.5
ILO unemployment rate	-	6.7	6.0	6.3	5.0

Sources: National Commission for Statistics (1998 and 1999b)

In December 1999 the highest unemployment rate of all Romanian counties was registered in Hunedoara County (20.4 %) and 4 Moldavian counties. On the other extreme we find Bucharest (5.8 %) and two north-western counties. Unemployment is particularly high among young people, making up 18.5 % of ILO unemployed. The age group structure was defined by groups who were differently related to work: children (0-14 years), young people (15-24 years), those of working age (25-59 years) and old people (60 years and over). The figures for each county show the dominance of the old in the southern part of the country (agricultural region), and the high numbers of children and young adults in northern Moldova (also an area with much agricultural, and the poorest region in the country). In southern Moldova and the Bărăgan Plain we find an alarming aspect: the proportions of children and the old are both above the national average, meaning that the workforce is relatively small and the number of economically dependent persons relatively large. In contrast, working-age groups dominate in the capital city.

Before 1989, most social insurance schemes were run by industrial enterprises and local administrations. However after 1990 companies ceased to provide such services or to act as job centres, and local public administrations were obliged to take over these functions. The incoherence and structural ambiguity of the system resulted in severe structural problems in industrial regions; laws were applied arbitrarily and there were problems in the handling of funds and the initiation of social programmes at a local level.

In 1998 the average number of doctors in the public sector was 1.84 per 1000 inhabitants. The county level indicator varies between 0.82 and 3.98. Bucharest has the best position, followed by counties with medical universities: Cluj, Timis and Iasi. The lack of doctors in industrialised cities and regions is not in fact related to the industrial character, but rather to living standards. Poor regions, whether they are industrial or agrarian, do not attract doctors.

Crime rates are rising, in line with general European trends. The share of crimes against private and public property has increased, as against a decreasing share of violent attacks (even if the absolute figures show increases). In 1998 the highest crime rates were in Vrancea, Suceava, Ialomiţa and Hunedoara counties, all areas suffering high levels of poverty, with only Hunedoara being highly industrialised.

One of the consequences of the transition is a reduced participation in education. With the exception of kindergarten services and higher education, there are negative tendencies at all other levels. The worst situation is found at the high school level (years 9-12), with major implications for the educational capital of the population. The reasons can be found in the financial difficulties of families, the general low priority currently assigned by some sections of the population to education, and the high percentage of unemployment among high school graduates.

After 1990, the migratory flows changed direction; in the last 40 years this was towards cities, contributing to urban growth with a percentage of 52 %. More recently the number of people who have been moving from rural areas to urban settlements has dropped by half, while urban-rural flows have tripled. In 1996, for the first time, the yearly flow to rural areas exceeded that to urban areas. The most affected cities are the ones having undergone forced industrialisation, which indicates that the out-migration from urban areas is not a sign of "suburbanisation" (wealthy people moving to luxury houses on city outskirts) but rather of "re-agrarisation".

1.2.4 Spatial Integration and Political Framework

Generally speaking, regional and urban-rural disparities in Romania have historical, geographical, economic and cultural roots. During the time of the communist regime, industrialisation was considered the best way to ensure a spatially balanced economic development and to diminish urban-rural disparities. The result was, however, to increase the gaps between rich and poor regions, as well as between villages and cities. The transition only served to deepen the long-standing disparities, and resources now tend to go where maximum turnover can be obtained in minimum time. The structural fragility of the economy in weaker regions is quite visible: the dependence on a single branch, the low degree of urbanisation and low attractiveness of artificially "swollen" cities, as well as the population's dominantly rural behaviour and cultural pattern.

There are eight Development Regions in Romania. They do not represent administrative units with limited autonomy, but constitute a framework for the conception, implementation and assessment of regional development policies, as well as providing the basis for statistical data-collection. The regional development policy is a component of Romania's National Development Plan (October 1999), the main purpose of which is to diminish the accumulated economic and social imbalances, to prevent new imbalances arising between regions and to support sustainable development of all regions in the country. It can be said that now a new approach to regional development has been undertaken, based on the principles of decentralisation, partnership and strategic planning. This enables the authorities and local and regional communities to get actively involved in the promotion of their own interests, and to initiate and promote regional development programmes and projects. Meanwhile the government is able to implement structural-type policies.

2 Specific Problems and Reasons for Decline

2.1 Specific Problems

Old industrialised cities and regions in Romania show several specific problems:

Economic problems: Regardless of whether socialist or pre-socialist, the general state of industrial cities and regions is that of decline. This implies both falling output and increasing unemployment. The adaptation problems encountered by enterprises are proportional to their scale.

Social problems: People's incomes and purchasing power has dropped in the past ten years, with poverty widening, especially in industrial cities and regions. This, on one hand, puts pressure on the social assistance budget, and on the other hand, reduces the entrepreneurial and investment possibilities inside the country. Furthermore, the unemployment rate has increased, and less favoured social groups have grown in number. The poverty and the incoherence of the public health sector's reform have led to the rise of various health problems in old industrialised cities.

Cultural problems: Fifty years of "socialism" has left its mark on people's ways of thinking. The overnight shift to a different economic and institutional framework led to various adaptation problems. High school and university graduates and those already in the workforce, all had to reconsider their social status. Suddenly people were required to take an active attitude towards work, to be self-reliant and to show competitive spirit. Many Romanians still have problems in understanding the new situation. Another challenge to traditional ways of thinking is the creation of a capitalist-type of entrepreneurial spirit. The first private firms that appeared after 1990 concentrated their efforts on making quick profits, without formulating medium-term development strategies and marketing policies or forming more efficient types of enterprises, such as holdings.

Ecological problems: One consequence of the use of old equipment and technologies was the high pollution level caused by local factories and enterprises. Today, many sources of industrial pollution have been eliminated, only to be replaced by new producers of diffuse pollution. Measures towards urban landscape protection in industrial cities face serious difficulties. For example, the lack of funds for reconversion, rebuilding and restoration, as well as the difficult regulation of construction are all negative factors.

"Image" problems: Many of the industrial regions facing structural changes have a great potential in other economic fields, such as tourism, services, forestry etc. The local population often recognizes these possibilities, but there are only weak (or non-existent) local resources for investment. Well-directed publicity campaigns could possibly attract investors from other regions or even from abroad. However such efforts currently remain unrealised, either from a lack of specialist knowledge or because of a more wide-spread misgiving about the effectiveness of such methods.

2.2 Reasons of Decline

The following four aspects concerning the reasons of decline of industrial cities and regions can be distinguished:

Global reasons for industrial decline are, firstly, the preservation of the old economic structures prior to 1989 and the slow pace of privatisation and restructuring; secondly, disagreement on the necessary steps for successful economic restructuring regarding monetary and banking policies (e.g. expensive and non-selective credits), especially when one takes account of the undercapitalised state of enterprises; and thirdly, the financial blockage (debts are not paid off) and high volume of company debts in the economy, which deter possible investors from viewing state-owned enterprises as attractive possibilities for privatisation.

"Post-socialist" reasons are the disappearance of the COMECON market and the decreasing demand for coal, metallurgic products, machinery, equipment and other such items on the internal and external market. Any long-term and strategic thinking and practical efforts at adaption were entirely lacking, so that difficult decisions were only postponed to a later date. Most state-owned enterprises showed low flexibility regarding the demands of a market economy (lack of correlation of the number of employees with decreasing output, or of the production profile with demand). After markets were opened, Romanian products proved to be of low international quality and expensive compared with those of competitors, leading to a drop in demand and, consequently, falling production. The interference of politicians in the management and decision-making of state-owned companies also proved highly counter-productive.

Branch-typical reasons are the low technical level of available industrial equipment and obsolete production technologies (energy-intensive, highly polluting, inefficient); the excessively large scale of enterprises; the growth of internal competition, with the appearance of new productive units (in the case of manufacturing industries); the lack of fiscal incentives for industrial manufacturing branches; the uncontrolled rise of energy and water prices for industrial customers; and finally the decrease of research, planning, and prospecting measures for further branch development.

Country-specific reasons can be found in the insufficient correlation of restructuring programmes with those for local and regional development and those for social security and professional retraining. Because of forced industrialisation, some enterprises are situated in poor locations. The legal and institutional framework for finance is unstable and unsatisfactory. Further country-specific reasons are: increasing bureaucracy; reduplications (the same problem is dealt with by two or more authorities), unclear responsibilities (no one feels responsible) and rivalries in decision-making; non-existent tariff policies for electricity and no regulation of the internal market concerning energy and raw materials, increasing the market price of Romania's products. The forex effort (payments in foreign exchange) is very high, because Romanian industry imports a large share of raw materials. The belated elaboration of a national strategy concerning mining and metallurgy, and unjustified rises in wages as compared with productivity (mining, energy), are also reasons for decline.

3 Strategies for Further Development

There are several strategic documents of importance for the development perspectives of old industrial cities and regions: the National Strategy for Sustainable Development (1998), Romania's Industrial Strategy and Policy (1996), the National Regional Development Plan (1999) and the Medium Term National Strategy of Economic Development (2000).

In the *National Strategy for Sustainable Development* (NSSD) (Guvernul României 1998), Romania's national long-term strategy is presented. This strategy is intended to cover the period from 2000 until 2020, and projects that by 2020 Romania's per capita GDP will be 50 % of the EU15 member-states' average. Romania's major goal is to reduce the economic gap separating it from the EU countries, and this is a condition for its accession to the EU. The main indicator taken to gauge a country's performance in relation to the EU region is GDP per capita. The NSSD indicates a scenario of sustainable development for Romania's process of convergence; according to this scenario, an average annual increase of GDP per capita of 6.5 % is necessary to ensure that Romania reach a share of about 50 % of the average EU level by the end of 2020 (the present level is 22 %). The NSSD is based on three major policies:

- conformation to the basic principles of the market economy;
- accession to the EU, in order to provide economic growth in the context of democracy;
- protection and sustainable improvement of the environment.

The overall objective of Romania's long-term strategy concerning industry is the improvement of competitiveness and achievement of steady and sustainable economic growth. This objective will be accomplished by accelerating and completing privatisation and restructuring processes (except for the natural resource monopolies, which will stay under public control); and by stimulating both foreign direct investment and a domestic capital market. At the same time, industrial reforms will take into account protection of the environment.

The following important operational objectives effecting the future of industrialised cities and regions are quoted from the *Romania's Industrial Strategy and Policy* (Ministerul Industriei şi Comerţului 1996):

- speeding up and completion of privatisation, a way to increase competitiveness and flexibility on the domestic and foreign market;
- support of the restructuring process in industrial branches that are still vigorous and competitive on the international market, improving the quality of products;
- liquidation or reconversion of loss-making industrial units with no future on the competitive market;
- promoting the industrial-financial concentration through associations of "holding" type, in order to increase the competitiveness;
- promoting industrial co-operation with international companies, in order to improve labour productivity, the competitiveness of products and to increase the value added;
- priority is given to supporting those industries that are export-oriented and/or provide high value added with high technology products;
- development of industrial services, both upward (research-design, technological engineering, testing etc.) and downward (distribution, logistics, advertising etc.);
- promoting sustainable industrial development within the framework of environmental protection, with more effective use of resources.

The *National Regional Development Plan* of Romania (Agenţia Naţională Pentru Dezvoltare Regională 1999) identifies two categories of strategic industries that will benefit from special support:

a) Basic industries – whose existence at national level must be assured. These include production, transport and delivery of electric and thermal power; operating, processing, transport and storage of oil and fuels; defence industries; the production of some pharmaceuticals and products needed for dealing with a national emergency.

b) Support industries – which support sustainable economic growth, provide employment, increase labour productivity and make efficient use of raw materials.

Romania's Medium Term National Strategy of Economic Development (MTNS) (Guvernul României 2000b), which was presented to the European Commission in March 2000, defines ways of overcoming the present poor state of the economy (characterised by weak institutional structures and economic agents, which show behaviour patterns at variance to the principles of a market economy).

The Jiu Valley – Perspectives after the Failure of a Strategy

The Jiu Valley area (954.3 km^2) comprises 6 urban centres and is situated in the central-western part of Romania, in Hunedoara County. It is the primary lignite producing area of Romania, providing about 90 % of the national coal output. During the Socialist era inter-regional migration meant that the population doubled, reaching a maximum of about 172,000 inhabitants, most of whom were dependent on mining. The regime managed to foster a strong professional pride among the miners with the understanding that they provided the economic basis of the country. After 1988 the area experienced a radical decline in coal production, falling to less than 50 % of the 1989 output. Eventually the downward slide was halted and reversed: in May 2000 production grew by 13 % compared to the level in May 1999. However, it should be noted that with increasing productivity the unemployment rate in 2000 was the highest in Romania (20.4 %). The area lost population both by natural decrease and by migration.

Government measures in 1997 led to a severe cut in subventions (from 45 % to 20 % of total costs), 8,000 miners were dismissed and the most unprofitable sites began to close. Those dismissed were offered 20 compensatory wages; this was supposed to provide some start capital for small private businesses. In addition, the government initiated programmes for retraining and professional counselling. The impact of these measures was not as intended. Because of the general lack of initiative shown by the unemployed and also in view of the insufficient technical support, the allocated funds were spent with no significant conversion taking place. In January 1999 a major social conflict was the result, requiring police intervention. The government set up programmes to create alternative skills and workplaces, but this met resistance from the miners: they displayed no willingness to be retrained, and were reluctant to relocate to find work. After the Law for Disadvantaged Areas (Monitorul Oficial al României 1999) was ratified, the Jiu Valley area was the first area to be declared "disadvantaged". This will lead to the creation of stable financial incentives for investors, with the condition that those previously dismissed are given jobs. Several projects intended to help ease the restructuring effort are underway, some with international funding: "Rapid impact measures for the creation of workplaces in the Jiu Valley and Gorj mining areas", for which Phare funds were allocated, comprises 28 projects; "Employment of workforce and social protection", financed by the World Bank, aiming for professional retraining, counselling and a redistribution of the workforce, and "Rehabilitation of the sewage system and wastewater treatment in the Jiu Valley", in the framework of the RICOP programme (Industrial Restructuring and Professional Reconversion), with funds from the EU and the European Bank for Reconstruction and Development. Another law expected to bring results in industrial restructuring and investment is the Law for Industrial Parks (2000). Considering the local topography and the fact that over 90 % of the area is forested, tourism and forestry in the Jiu Valley have strong potentials for development.

Dorottya Pantea, Urbanproiect, Bucharest

This document is accompanied by an Action Plan for the Government Programme for 2000 and the period 2001-2004 (Guvernul României 2000a), that proposes concrete measures with a detailed timetable. The MTNS is the first strategic document that is expected to be respected by different governments, no matter what their political orientation. MTNS proposes, among other things, a set of structural adjustment and economic development policies, setting out in an integrated way the specific problems and future fields of action. Unfortunately, no spatial considerations are included. Regional and local authorities have to adapt and submit to the nation-wide regulations.

The sectoral policy for industry has the following objectives: extending the process of redesigning industrial capacity and structure, stimulating potentially competitive big and medium-sized enterprises, and the encouragement of co-operation with EU partners. In addition attention is paid to the selection and re-scaling of economic agents in order to ensure a proper business environment. Privatisation of industrial companies has to be completed and state-owned industrial units should be exposed to market forces, with a strengthening of the compliance with the laws regarding competition. Other objectives for the sectoral policy for industry are the restructuring of energy and material intensive sectors, and of those with under-utilised capacities; the efficient development of internal Research and Development (R&D), including micro-economic R&D; the promotion of strategic alliances, holdings and group companies; rapid development, outsourcing and specialisation of production-related services. And last but not least, exports of high value-added products must be encouraged, as well as support given for the global expansion of high-tech industries.

Based on the MTNS, an Action Plan for 2000 and the period 2001-2004 was issued in June 2000, outlining specific measures towards the structuring and recovery of the national economy. Industry and industrial regions are approached in an integrated way in several chapters of this Action Plan. The main proposals and actions referring to, or having an impact on, industrialised cities and regions, are presented here:

1. Financial assistance for public companies should cease; sound delimitation of the fields of intervention of the state in industry is necessary (only natural resource monopolies)
2. The state-owned sector should be reduced, by involving investment banks and through contracts with strategic investors;
3. Barriers that hamper the breaking-up of large companies should be removed (payment or conversion of budget obligations, protection of new owners against the old ones);
4. The system of debt compensation should be extended to all economic agents, through inclusion of obligations to the state budget, the social assistance budget and local taxes; this will reduce financial blockage;
5. The procedures for founding an industrial enterprise should be simplified;
6. Competition in the utilities sector should be promoted by the privatisation of electricity, gas, thermal energy distribution and railway companies;

7. Enterprises with debts exceeding 85 % of assets should be liquidated; other agents implementing viable restructuring programmes should be supported.

In order to promote the regional development policy, the Government, in co-operation with local and county authorities, has finalised the legal framework and the institutional structures at central and regional level necessary to implement the specific measures. Laws have also been drafted intended to animate the entrepreneurial climate, to encourage direct investment and establish free trade areas.

An important component of the regional development policy aims to support disadvantaged areas with severe structural problems (mono-industrial structures, high unemployment) by granting important incentives to investors in these areas, on the basis of already approved legal regulations. There are 25 disadvantaged areas in Romania. The main investment incentives offered in these regions are: exemption from customs taxes, profit taxes and/or from value added tax (VAT) for imported raw materials and equipment, and grants towards such things as export stimulation. In 1999, several programmes were introduced for regional restructuring, retraining and professional reconversion, as well as favourable credits for SMEs in the disadvantaged areas.

References

Agenţia Naţională Pentru Dezvoltare Regională (1999) Romania's National Regional Development Plan, Bucureşti

Chiribucă D, Comşa M, Dâncu VS, Rotaru T (2000) The social effects of economic restructuring of mono-industrial areas – regional study for the Jiu Valley, no publisher and place of publication specified

Cucu SV (1996) Economic geography of Romania, Editura Glasul Bucovinei, Iaşi

Editura Expert (1998 and 1999) Romania's National Human Development Reports, Bucureşti

Ministerul Industriei şi Comerţului (1996) The Industrial Strategy and Policy of Romania, Bucureşti

Monitorul Oficial al României (1998) Law for Regional Development in Romania 151/1998, Bucureşti

Monitorul Oficial al României (1999) Law for the Establishment of Disadvantaged Areas, 20/1999, Bucureşti

National Commission for Statistics (1991-1999) Romanian Statistical Yearbooks, Bucureşti

National Commission for Statistics (1999b) CESTAT – Statistical Bulletins – collection, Bucureşti

Guvernul României (1998) National Strategy for Sustainable Development of Romania, Bucureşti

Guvernul României (2000a) Action Plan for the Government Programme for 2000 and the period 2001-2004, Bucureşti

Guvernul României (2000b) Medium Term National Strategy of Economic Development, Bucureşti

Slovakia: Transforming Industrial Regions and Preparing for EU Regional Policy

Maroš Finka[1], Robert Janaček[2] and Dagmar Petríková[1]

[1] Slovak Technical University, Faculty of Architecture, Bratislava,
[2] AUREX Ltd., Bratislava

1 History of Industrial Development in Slovakia

The first stage of industrial production, primarily the extraction of raw materials and the production of consumer goods, dates back to the Middle Ages. Industrial development was stimulated by the arrival of German colonists in the 12th and 14th centuries; by the 16th and 17th centuries mining and metallurgy had expanded under the influence of these German immigrants, and Slovakia became the industrial powerhouse of the Hungarian empire. The industrial revolution arrived in the second half of the 19th century, later than in Bohemia and other developed European countries. The bourgeois revolution of 1848/1849 and the Austro-Hungarian influence since 1867 led to political acceptance of capitalist industrialisation. In 1857 about 70,000 people were working in crafts and industry (11 % of the economically active population); this number increased to 260,000 (20.9 %) by 1910. At the beginning of the 20th century industry was marked by a low degree of concentration. The Czechoslovak Republic was founded after World War I; at the time, Slovakia's industry was the most developed in the Hungarian part of the former monarchy. Subsequently industry went into decline, deepening territorial disparities of industrial localisation in Czechoslovakia in the interwar period.

The socialist industrialisation of Slovakia after World War II was one of the main features influencing economic development as well as affecting all other spheres of social development. From 1948, industrial production increased dramatically, expanding by a factor of 29.9 by 1985. Within this general economic expansion the production of manufacturing tools increased 2.4 times quicker than the production of consumer goods.

The most successful industrial regions were the large regions of Middle Váh Valley (along the river Váh, centred around Trenčiansky kraj), Bratislava and Košice (cf. map), with twice as much basic industry than the average for all regions.

SLOVAKI
CITIES, ROAD
AND KRAJE
Town or City
Capital
Road or Mot
Border of Kr
(NUTS 3)
River
Lake
0
25
50
Scale 1:2800000
Sources:
ESRI ArcData,
Basemap 1999
Map: Lintz, Witschas
Leibniz Institute
POLAND
CZECH REPUBLIC
HUNGARY
UKR.
AUS.
Wisla
Vltava
Odra
Vah
Hernad
Tisza
Danube
ZILINSKY KRAJ
PRESOVSKY KRAJ
KOSICKY KRAJ
TRENCIANSKY KRAJ
BANSKOBYSTRICKY KRAJ
TRNAVSKY KRAJ
NITRIANSKY KRAJ
BRATI-SLAVA
Zilina
Martin
Poprad
Presov
Kosice
Trencin
Prievidza
Banska Bystrica
Zvolen
Lucenec
Nitra
Trnava
Bratislava
Komarno

These three regions have the highest values in industrial production. The Middle Váh Valley boasted an industrial workforce of about 400,000, more than 3 times the regional average, and the regions of Bratislava, Upper Nitra and Lower Váh could claim similar figures. The industrial regions of Middle and Upper Váh Valley had a special position as core regions, affecting the development of industry in the adjacent territories of Orava, Kysuce and Turiec.

In 1980 more than one million people (19 % of the total population of Slovakia) lived in the two largest regions, Bratislava and Middle Váh Valley. The degree of concentration reflects primarily the suitability or unsuitability for industry localisations. The historical evolution of the country's industrial sectors and certain favourable factors have ensured that the largest share of industry is localised in the lowlands in the north-west of Slovakia (comprising 16.5 % of the country).

"In 1980 the number of settlements with industrial production was 411 units, with 168 units in the category of industrial centres and 243 industrial localities" (Mládek 1990). Most industrial centres and localities in Slovakia showed a highly specialised structure, with one or two industrial branches being dominant. Only the larger industrial centres developed a multi-branch industrial structure. Localisations and the structure of industry were determined by state policy. This focused on:

- the creation of jobs (without gender bias) throughout the country;
- the stabilisation of a politically-acceptable social structure (percentage of workers);
- securing self-sufficiency in all branches of goods production;
- export to the states of the COMECON;
- providing special production (armaments) for the states of the Warsaw Pact, and the strategic localisation of such production.

The issues of environmental protection, diversity of industrial structure, competitiveness and the efficient use of local potential as the basis for sustainable development were of secondary importance.

2 Industry and Industrial Areas after 1989

The year 1990 saw a break in the development of industry. Decisive political changes led to a considerable change in the economic orientation. Previously, Slovakian industry had been highly dependent on the import of raw materials and was oriented towards the markets of the COMECON. The country had a low innovative level of production, low competitiveness on the EU markets, and a narrow range of production (arms industry). All these factors determined the need for the restructuring of industry.

17 basic industrial regions could be identified in 1990. According to the evaluation in "Teritorialne priemyslené útvary Slovenska" (Mládek 1990) the first category included the 5 largest regions – the regions of Middle Váh Valley, Bratislava, Košice, Lower Váh and Upper Nitra. In 1980 they concentrated 60 % of production value, 50 % of employment and 55 % of the value of capital stock of the for-

mer SSR's industrial sector. The regions of Upper Hron, Lower Nitra, Upper Váh, Spiš, Humenné, Záhorie, Lučenec, Šariš and Middle Hron are characterised as average regions according to their size. The regions of Michalovce, Lower Hron and Gemer are the smallest industrial regions – in 1980 their total potential only made up 8 to 9 % of Slovak industry.

In 1980 the highest rank for industrial specialisation was accorded the regions of Middle Hron and Košice, with the highest variety of industrial structure found in the regions of Middle Váh Valley, Lower Váh Valley, Upper Hron, and, particularly, in the region of Spiš.

The process of contracting public and private investment that began in 1990 was halted in 1994, and GDP increased by 4.9 % in that year. From 1995 a steady deceleration of growth in GDP can be recognized. The share of industry (including construction) in GDP also changed considerably, falling from 61 % in 1986 to only 26.6 % in 1998. The regional breakdown of added value in 1997 is shown in Table 1 (cf. also Map).[1]

Table 1. Breakdown of added value in 1997 in millions of Slovak crowns, (1 USD = 45 Sk)

Region (Kraj, NUTS 3)	Added Value	
	Mil. Sk	%
Bratislava	189,696	32.3
Trnava	53,848	9.2
Trencín	57,723	9.8
Nitra	55,201	9.5
Zilina	52,451	8.9
Banská Bystrica	59,310	10.1
Prešov	45,217	7.7
Košice	73,084	12.5
Total SR	586,530	100.0

Source: Statistical Office (Štatistický úrad SR), 1998

The leading positions of the capital Bratislava and the eastern Slovakian city of Košice are defining characteristics of the economic structure and potential of the regions in the Slovak Republic. Below the regional level come the districts, showing considerably developed urban centres such as as Žilina, Banská Bystrica, Prešov, Trnava, Trenčín and Nitra. The main development axes of Bratislava–Žilina–Košice and Bratislava–Nitra–Banská Bystrica, are clearly dominant. The leading position of Bratislava as the capital is bolstered by its location on the borders of the European Union, by a new orientation in foreign trade as well as by its regional background with a highly qualified and skilled labour force. The region of Bratislava is responsible for more than one third of the Gross Domestic Product (293,000 SK / 6,500 US$ per person). This means that the region produces more

[1] From 1996-2003 the territorial administrative structure comprised 8 regions on NUTS 3 level (kraj) and 79 districts on NUTS 4 level (okresy). In 2002, after regional elections, the regions became self-governing units; from 2004 a new administrative structure of 50 administrative districts (obvody) will be introduced to the regions.

than three times the average per capita GDP of the Slovak Republic, as seen in Table 2.

Table 2. Gross Domestic Product in 1997 (1USD = 45 Sk)

NUTS II	Regions (NUTS III, Kraj)	Share in GDP (%)	GDP per capita (Sk)	Unemployment (%)	Investments (% SR)
Bratislava	Bratislava	35.0	292,900	4.9	52.1
Western Slovakia	Trnava Trenčín Nitra	25.4	70,070	12.3	20.9
Middle Slovakia	Žilina Banská Bystrica	19.3	73,975	13.6	13.8
Eastern Slovakia	Prešov Košice	20.3	68,880	18.3	13.2
Average		25.0	96,500	13.4	25.0
Total SR		100.0	96,500	13.4	100.0

Source: Buček 1998

At present more than 60 % of all foreign capital investment is concentrated in Bratislava (Volkswagen, Tesco, Volksbank and others), with 18.5 % in the districts Žiar nad Hronom, Poprad, Prievidza, Humenné and Nitra; 43.4 % of foreign investment has been in industry, 32.4 % in trade and 15 % in banking and insurance.

The majority of exports are products highly sensitive to external fluctuations in prices and boom patterns in foreign markets. From a regional point of view changes in the structure of exports are significant in so far as exporters are the key regional employers. In the past the failure of exports was felt in the region of Middle Váh Valley (particularly in the towns Dubnica and Považská Bystrica) where weapons were produced. The region suffered from the break-up of the Warsaw Pact and the political decision to halt the export of arms. At present this is still a serious problem in the region of Košice and the VSŽ Steel Works; 11 % of the Slovak Republic's exports are produced in this area.

The German Institute EMPIRICA (Bonn, 1993) analysed 414 European regions regarding their potential to become magnets for investment. The localities with the highest potential were divided into three groups. Bratislava was placed in the group of most favourable localities, because, as the authors say, "in addition to being an economic and administrative metropolitan region, it has an excellent background of research". The study also designated a cross-border area comprising three neighbouring regions of Slovakia (Bratislava), Hungary (Györ), and Austria (Burgenland) as the "Golden Triangle of Production".

3 State of Industrialised Cities and Regions Undergoing Restructuring

3.1 Reasons for Decline

When analysing the problems of industrialised cities and regions in the Slovak Republic the factors initiating structural change must first be identified. Even if it is clear that basic structural change resulted from upheavals in society and shifts in the economic foundations, still these changes can have various specific associated and subsidiary causes, with diverse external symptoms and consequences that require remediation.

The basic characteristics of the transformation processes in industrialised cities regions can be defined as follows:

- rapid post-war industrialisation, with simultaneous industrialisation and urbanisation processes, a tradition of narrow industrial production (e.g. arms production);
- isolated branches of industry with highly specialised labour force, existence of relatively compact industrial complexes with low economic diversification, high regional dependency on a few dominant industrial sectors;
- production oriented exclusively towards the states of the Eastern bloc and the Soviet Union, dependence on the states of the COMECON for raw materials, low competitiveness on the world market because of the isolated nature of COMECON, collapse of these former markets;
- specific natural and socio-political features: geographical position in Europe, valuable natural areas, dependency of industry on imported raw materials, precious historical and cultural heritage;
- combination of industrial transformation with general social transformation, division of the Czech and Slovak Federal Republic (CSFR) and the processes of EU accession;

The general processes of economic transformation were characterised by the establishment of free-market institutions, the encouragement of entrepreneurial activities, and a reduction in the significance of state ownership by part privatisation and the injection of foreign capital. These changes brought an increase in social and regional disparities.

These factors have triggered structural change in industry and determined the problems of industrialised cities and regions in the Slovak Republic. Further impacts have been the global transformation from an industrial society to an information society and the pressures to ensure sustainable development (nuclear energy development).

The following aspects must also be taken into account: particular features of the transformation process in relation to their spatial impacts (e.g. geographical position, mechanisms of distribution, access to markets), and problems and dangers of other uncontrolled processes (social, ecological and spatial processes). Structural changes in the Slovak Republic take various forms and have various effects:

- shift of economic focus from industry to tertiary/quartery sectors;
- changes in industrial production from mass-production and high consumption of energy and raw materials towards flexible production in small specialised firms, with lower energy and raw material demands;
- changes in the relation of labour to workplace;
- changes in the significance of location factors, from hard to soft factors.

3.2 Evaluation and Typology of Industrialised Cities and Regions after 1989

A current evaluation of the Slovak industrial regions relies on data from a 1998 study prepared by the Office for the Strategic Development of Society, Science and Technology of the SR. A variety of indicators are used to evaluate the country's 79 districts, newly established in 1996. For every indicator and every region (and district) a point value was given (with appropriate weighting), corresponding to the rank of the region (district) within the Slovak Republic.

The *production performance* of the regions was defined as the first criterion. The data used for the classification were the statistical data for the first half of 1997, with indicators chosen as follows:

- production of industrial goods,
- volume of construction production,
- retail sales, and
- other market services.

As supplementary data the study used the average monthly salaries and investments per capita. The highest per capita production was found in the regions of Bratislava and Košice, being the regions with the highest concentration of industrial production and know-how. The smallest per capita production was found in the districts Sobrance, Šariš and Gelnica, in the region Spiš; the latter is one of the industrial regions worst affected by structural change in mining and machine construction. The best production performances were achieved by districts in the regions Bratislava and Košice; the worst by districts in Michalovce and Šariš.

To evaluate the *labour market* the study used an indicator combining:

- the number of workers,
- the unemployment rate,
- the number of legal and physical entities in a given district or region per 100 and 1,000 inhabitants.

The number of workers per 100 inhabitants was highest in the districts of Bratislava and lowest in the districts of the Prešov region. The lowest unemployment rates were recorded in the districts of Bratislava (3.6-4.5 %) and in the Middle Váh Valley (4.3 %). An unemployment rate of more than 20 % was recorded in the smallest industrial regions of eastern and southern Slovakia (Gemer, Michalovce). One of the important indicators used in this analysis was a quotient comparing the labour market conditions in the various districts and regions of the Slovak Republic. The best labour market conditions were identified in the regions Bratislava, Middle Váh Valley and Upper Hron (district Banská Bystrica). The most serious labour market problems were found in the regions of north-eastern Slovakia (region Spiš, Šariš, Humenné) and eastern Slovakia, with the exception of Košice district.

Another important indicator for the health of the economy is the condition of the *natural environment* (forming one localisation factor). The negative environmental consequences of economic activities in regions are registered by this indicator. The analysis of environmental conditions was based on the Environmental Report of the SR for the year 1996. However this only gives a rough picture of the state of the natural environment in analysed regions. The best results from the old industrialised regions were found in the Middle Váh Valley region (districts Púchov – good environmental standards, Bytča and Považská Bystrica) and the regions of north-eastern Slovakia (unsatisfactory environmental standards). The poorest results were recorded in the Bratislava region – with a damaged environment (districts of Bratislava I, II) and in the region of Košice, with a heavily damaged environment.

The indicators formed by *demographic characteristics* determine the socioeconomic state of the old industrial regions undergoing structural change. They include the following: a population ageing index, an index for the economic load of the population, as well as indicators for the natural growth of the population and migration patterns. Values for the population ageing index, calculated by multiplying the quotient – productive population over unproductive population – by 100, were highest in districts outside the old industrial regions. The most unfavourable values were found in the districts of the Bratislava and Nitra regions. The index value calculated as non-productive (post-productive + pre-productive) population divided by the productive population (and multiplied by 100) was highest in the districts of Košice and in the region of Bratislava. The highest natural population growth per 1,000 inhabitants was recorded in the northern and north-eastern regions, while the largest population decrease was seen in the region of Bratislava. The most unfavourable values were in the districts of north-eastern Slovakia. Migration per 1,000 inhabitants was highest in Bratislava region. The lowest migration was recorded in districts in the Middle Váh Valley and in the north-eastern regions.

The highest level of *technical infrastructure* was found in the Bratislava region, with the lowest level in the districts of the north-eastern regions. The highest level of education was found in districts of the regions Bratislava, Košice, Banská Bystrica and Zvolen. The north-eastern regions had lower levels of education.

The *overall position* of a district and region within the SR is characterised by a complex indicator. The more developed were determined to be districts in the regions Bratislava, Upper Hron (Banská Bystrica) and Middle Váh Valley (Trenčín). The lowest level of socio-economic development was identified in the districts of eastern and north-eastern Slovakia (with the exception of Prešov and Košice districts) and in the region of Nitra.

One conclusion from this analysis is that the industrial cities and regions primarily suffering the negative consequences of structural change are those whose economic base was less diversified, being tied to some dominant branch of industry or to low-skill enterprises. This is shown by the poor condition of districts in north-east and south Slovakia. In the most dynamic regions of Bratislava, Middle Váh Valley and Košice, industry and the underlying economic structures are undergoing restructuring processes, with new emphasis on tertiary and even quartery sectors.

4 Strategies for the Development of Industrialised Cities and Regions[2]

4.1 Overview

Slovak regional policy includes strategies for the revitalisation of regions undergoing structural change. The National Plan for Regional Development passed in 2001 is a basic document for regional development and the coordination of structural instruments. The following five priorities have been determined to ease the current situation in regions undergoing structural changes: formulation of a regional development policy; establishment and strengthening of regional development institutions; deepening of co-operation between authorities and institutions responsible for different aspects of regional development, including cross-border co-operation to resolve issues of land planning; steady application of EU principles and procedures for the preparation and management of regional development projects. These priorities, gathered under the title "Establishment of a Complex Regional Development Policy", are also decisive for the National Programme to adopt EU Legislation, in the Sectoral Operational Policy at regional level and in the co-ordination of structural instruments. The elaboration of the strategy was based on the evaluation of socioeconomic conditions and the typology of districts and regions in the Slovak Republic, as well as by considering the strategic intentions and objectives corresponding to the objectives and procedures of structural policy in EU countries. At the state level the logical structure of this approach included the following steps:

a) formulation of long-term goals;
b) designation of medium-term tasks and objectives, the accomplishment of which would help to ensure wider strategic goals;
c) defining priorities and measures, such as:

- concrete measures leading to the accomplishment of the set regional objectives,
- priorities in determining regional projects, and

[2] This chapter is based on Rajčák (1999).

The Region of Stredné Považie (Middle Váh Valley) – SWOT Analysis

One of the Slovakian industrial regions to have undergone important changes in industrial structure is the Middle Váh Valley in Trenčiansky region, situated in the west of Slovakia. In 2003 Trenčiansky region had 605,882 inhabitants (11.26 % of the national total). The main phase of industrialisation there began after WW II, with key factors in development being the strategic position of the region, its sufficient energy resources, substantial labour force, good transport links and substantial deposits of raw materials. Before 1989 the armament industry was prominent in the industrial profile of the region. Regional industrial development suffered in the years following 1989. Weaknesses that had been neglected for decades, such as an unhealthy bias towards heavy machinery, a low level of competitiveness as compared to Western industry, obsolete technologies and insufficient modernisation, as well as the absence of regional politics and market-economy laws etc., seriously undermined potential development.

Analyses of the strengths, weaknesses, opportunities and threats (SWOT) for Middle Váh Valley (MVV) were carried out by Robert Janacek (as a part of the FOCUS Project underlying this book) and Milan Rajčák (1999), and their conclusions form the basis for any recovery strategy. The most significant findings are as follows:

Strengths: diversified industrial branch structure, new and existing industrial businesses with the potential for expansion, well qualified workforce with good education level, rising average industrial monthly wages in MVV region, potential for establishment of new industrial and high-tech parks, regional development strategies.

Weaknesses: unfinished conversion of armament industry, production of basic goods with low share of finished goods, falling economic activity, serious structural problems, high level of business failures, insufficient resources, lack of industrial diversification forming risk factor for social stability, bad payment behaviour, unclear property rights.

Opportunities: geographical location can attract foreign investment and high-technologies to MVV region, EU support programmes and structural funds, new investment to existing producers having a positive influence on industrial development, establishment of scientific-research base for ongoing transfer of technologies between science and industry, establishment of business incubators for incipient entrepreneurs.

Threats: deepening of negative trends in industry, with negative effects for employment and development of individual branches, lack of publicity for business opportunities, slow-down in financial restructuring of enterprises, weak financing sector.

Robert Janaček and Matej Jaššo, Faculty of Architecture, Slovak University of Technology, Bratislava

- fixing the orientation of sector development programmes according to the social and economic needs of the territorial units;

d) monitoring and up-dating, implemented in close connection with approved concepts of regional economic and social development as well as socioeconomic development programmes in the districts.

Slovak regional policy pursues two strategic goals: supporting Slovak regions and increasing national prosperity, as well as improving institutional and administrative capacities for regional policy.

4.2 Strategic Intention 1: Supporting Slovak Regions and Increasing National Prosperity

Among the regions undergoing structural change a variety of internal development conditions and openness to integration processes can be found. One goal is to view regions as structural units in the national economic system, with the development of these territorial units raising the prosperity of the national economy. The Strategic Intention 1 has four objectives, specified in the following section. Objective 2 is the most important in the context of developing industrialised cities and regions.

Strategic objective 1: Reduction of disparities between regions and increasing their competitive strength. Socioeconomic change in the transformation period significantly increased differences in living and working conditions and potentials for development between Slovak's regions. Based on experiences of developed European countries, the aim here is to create a system of regional policy which minimises the negative spatial impact of structural change in order to support the long-term competitive strengths of individual regions. This will help support inhabitants facing weak starting conditions because of the low development potential, poor infrastructure and poor public services of their region. Priority will be given to districts lagging behind, as well as development poles and border regions. The following measures are foreseen:

a) implementation of structures for the development of regions in central Slovakia; this is a prospective development pole according to the government resolution no.: 315/1996 "Project Design for the Development of the Infrastructural Space of the Central-Slovak Region", with an urban agglomeration consisting of the two centres Banská Bystrica and Zvolen;

b) preparation of pilot projects for the introduction of EU structural policy to the region Banská Bystrica, as a gradual implementation of procedures and mechanisms for utilising EU structural funds. The project may be implemented within a Special Preparatory Programme for the employment of structural funds in candidate countries;

c) implementation of programmes within PHARE CBC, CREDO and Interreg IIC which support cross-border and inter-regional co-operation.

Strategic objective 2: Conversion of regions severely affected by industrial decline and structural crisis. The processes of transformation and structural reform have divided industrial companies and regions into several groups, according to how well they have managed to reorientate their production and manufacturing philosophy to the changed macroeconomic conditions and new market demands. The inequity and asymmetry of regional transformation was caused by the exhaustion of existing capacities and the slow pace in creating new production/export capacities. This in turn was a result of insufficient microeconomic adaptation and the small range of efficient structural changes. The necessity for structural change is most significant in monostructural industrial regions. Priorities are:

- easing the difficulties of non-adapted companies who have not been able to compensate a loss of former markets;
- loosening the restrictions surrounding bankruptcy proceedings in order to free the resources bound to struggling companies;
- resolving the problems of armament manufacturing companies;
- prevention of mining activities in unprofitable areas;
- bringing small and medium-sized enterprises with advanced scientific-technological production into joint innovative activities with other economic subjects.

In order to support the conversion of regions severely affected by industrial decline and structural crisis the following measures should be applied:

a) Support programmes of the Slovak Ministry of Economy to enhance the competitive ability of industry: programmes of technological and innovative development, improvement of quality and industrial design, export of Slovakian investment complexes and goods, the utilisation of risk-capital, the development of human resources, programmes supporting environmental management and environmental audits in industry, use of recycled raw-materials, development of the car manufacturing industry, programmes supporting the development of the engineering and armament industry in the group of DMD Holding a.s. Trenčín, development of industry aimed at the efficient use and processing of lumbar, development of utilisation of bio-technologies.

b) Tourism support programme, securing the growth of accommodation, catering trade and supplementary services, especially in localities affected by a decline of the engineering industry and by rising unemployment.

Strategic objective 3: Raising the employment rate and easing labour market problems in the regions. The disparities between the recorded unemployment rate of regions are considerable. To reduce these disparities co-operation between central authorities, business subjects and regional authorities is necessary. Greater financial support for an active labour market will alone not be able to check the high unemployment rate. Therefore the problems of an active employment policy shift from a macroeconomic to a regional and local level. Regional problems must be addressed, such as tendencies towards high unemployment and an ageing productive population in declining branches. Long term priorities for objective 3 are:

- support for the setting up of small enterprises and small businesses, with additional aid for employment in those enterprises and small businesses;

- elaboration and implementation of regional programmes for employment support and revitalisation; regional retraining programmes, regional retraining programmes for low-skilled workers and projects to bring the long-time unemployed back onto the job market;
- employment support for enterprises involved in revitalisation programmes;
- prevention of involuntary unemployment.

Short-term priorities for objective 3 are:

- creation of new long-term jobs in districts where the annual unemployment rate exceeds 25 %;
- preservation of jobs in healthy industrial branches and agriculture, forestry, and railway transportation;
- finding jobs for the registered unemployed, especially families where both partners are unemployed;
- preservation of existing jobs in firms which are key job providers (in areas with the highest unemployment rates);
- retraining employees in organisations which are implementing the restructuring of production or up-dating their manufacturing processes; also the targeted retraining of registered unemployed to meet regional demand on the labour market;
- creation of public-sector jobs in districts with insufficient entrepreneurial activities; employment projects with a special emphasis on the long-term unemployed.

Objective 3 will be achieved by implementing the following measures:

a) Creation and implementation of employment projects (including retraining, work rehabilitation, in particular for special target groups of the long-term unemployed and persons with restricted capacities).

b) Elaboration of a proposal for the amendment of the Slovak National Council Act No.: 387/1996 – Employment Act in its updated form.

c) Establishment of procedures for the merging of financial means for regional employment support and revitalisation programmes as financed by the National Labour Office, the National Agency for the Development of Small and Medium Size Enterprises, the Slovak Revitalisation Agency, and other organisations.

Strategic objective 4: Development of rural areas. Special care and protection must be given to rural areas as the areas most affected by structural change in agriculture and industry. During the period of industrialisation between 1945 and 1998 the rural population became dependent on income from non-agricultural activities. The development of new living conditions, work preferences and priorities led to a deterioration in the residential structure. Nonetheless, rural areas and their resident populations still represent considerable untapped resources for economic development. Slovakia is not an urbanised country, with 43 % of the population living in rural areas, 44 % in transition areas and 13 % in urban areas. Priorities in the development of rural areas are:

- supporting the preservation of agriculture in mountainous areas and other disadvantaged areas as a basic prerequisite for the good development of the land-

scapes, for ecological and social functions and for the preservation of the rural settlements;
- general economic development and increased employment in rural areas;
- natural resources utilised in a rational way, following the principles of environmental protection, keeping rural settlements intact;
- developing conditions for rural tourism;
- preserving the national and historical heritage, with development of the cultural and intellectual potential of rural areas.

The development of rural areas is supported by the following measures:

a) The PHARE Programme:
- Countryside Development Programme, based on a memorandum on the recognition and adoption of rules for rural development – including five pilot rural development projects;
- Support given to projects of regional and rural development within the Rural Development Fund.

b) Elaboration of a SR rural development concept in order to formulate strategic and specific objectives for rural development, outlining the principles of application of rural policy and specific proposals and measures.

c) Support for diversification of the rural economy and for creating added value in traditional and alternative (ecologically sound) agriculture, the processing industry, food industry, wood processing and other industrial branches.

d) Support for mountainous and other disadvantaged areas:
- by establishing a regional classification in the sense of EC Directive No 950/97 (Directive on the improvement of the performance of the agriculture structure);
- by creating an information and monitoring system to secure support; establishing a complex model-project to support mountainous and other disadvantaged areas;
- by founding projects of regional support with pre-accession help.

e) Development of rural tourism and local services for citizens.

f) Implementation of agro-environmental pilot and demonstration projects (including the securing of alternative economic uses of agricultural soil).

g) Completion of land ownership registration (as a basis for land market development and the establishment of institutions for the implementation of land modifications).

4.3 Strategic Intention 2: Improvement of Institutional and Administrative Capacity for Regional Policy

The economic and social coherence of cities and regions undergoing structural changes is a priority for EU enlargement. The implementation of a complex regional policy integrated with structural instruments requires the establishment or improvement of the appropriate institutional and administrative capacity. Only such a capacity will enable Slovakia to use EU structural funds successfully. In detail this strategic aim comprises four strategic objectives.

Strategic Objective 1: Formulation of a complex integrated policy of regional development, to create institutions of regional development and a corresponding legal framework in the Slovak Republic. Based on the experience of gradual regional development in Slovakia and also drawing on the experiences of developed European countries, it is necessary to create and strengthen the institutions and mechanisms of complex regional development, with special emphasis on the revitalisation of regions affected by structural change. Based on the new territorial and administrative division of the Slovak Republic, institutions of regional state administration and self-government should be created on a regional level. They will be gradually supplemented by organisations such as regional development agencies, funds and other actors which support the aims of regional policy. At the same time, in order to determine the conditions and rules for creating and implementing regional policy it is necessary to establish a legislative form for the co-ordination of the state administration and self-governing bodies. Priorities are the improved co-operation between authorities and institutions responsible for different aspects of regional development and emphasis on the benefits of regional development/research processes.

Objective 1 will be achieved by the following measures:

a) The creation of the National Agency for Regional Development in the Slovak Republic (NARR SR). This agency should secure economies in the use of resources allocated for regional development, especially funds from the state budget and from extra-budgetary sources, as well as foreign aid.

b) The gradual establishment of regional development agencies. Their main task will be – along with other bodies involved in implementing regional policy – to initiate and support the creation and running of particular development projects.

c) The preparation and implementation of concepts for economic and social development in the regions. These will be prepared by regional offices, based on the approved strategy of regional development. At this stage it is important to approximate to EU documents – in structure as well as content.

d) The preparation of programmes for the social and economic development of districts. These will be prepared by district offices. Their intention and content will be similar to regional development concepts.

e) Establishment of a research/development unit to deal with problems of regional development. This unit will solve the theoretical and practical problems of regional policy at state level, and the regional development and spatial planning difficulties of particular Slovak regions. It will prepare regional studies for the EC, the OECD, the World Bank etc.

f) Preparation and approval of the draft Act of the Slovak National Council on Regional Development. The purpose of this Act will be to fix the conditions and rules for the creation and implementation of regional policy and for the performance of the corresponding tasks of the state administration authorities and the self-government authorities. These latter will be the integrating bodies in creating and introducing complex regional policy, co-ordinated by structural instruments and compatible with the EU.

g) Creation of a system of co-ordination and information for the flow and use of means given to regions and towns/villages by state funds and the Fund of National Property.

h) Preparation and approval of the draft Act of the Slovak National Council on the establishment of a National Fund for regional development within the National Agency for Regional Development of the Slovak Republic. Its purpose is to determine the way of allocating subsidies and to ensure well-functioning institutions with good management and control of the financial means of structural policy – both within the Slovak Republic and the EU.

Strategic Objective 2: Creation and strengthening of budgetary and financial control institutions and mechanisms for regional development and structural instruments compatible with the EU. Slovakia has the long-term goal of joining the EU, dependent on the meeting of various conditions. The document "Partnership for Membership" gives one priority as the development of a complex integrated policy of regional development. Section 21 (Regional policy and co-ordination of structural instruments) of the "National programme for Acceptance of the Acquis Communautaire" has been developed from this priority. Preconditions for the realisation of these aims and the successful implementation of the Acquis Communautaire are the systematic professional training of staff in regional authorities and institutions, the comprehension of EU structural policy and the interconnection with the PHARE Programme.

Priorities are the gradual introduction of EU principles and procedures in the preparation and management of regional development projects, as well as the implementation of structural changes necessary to adapt to scientific/research capacities in EU countries.

The following measures will help to ensure the success of objective 2:

a) The selection of co-ordinated structural instruments of regional policy compatible with EU instruments and the establishment of procedures to optimise the employment of EU structural funds. This will create a complex strategy, allowing increasing convergence of Slovakian industrial structures with those within the EU;

b) The completion of projects within the PHARE Programme for Regional Development;

c) The establishment of regional and local employment projects according to the procedures of EU structural policy;

d) Involvement of the Slovak Republic in the special programme PHARE to prepare the way for EU structural policy (Finka 2000, 278).

5 Conclusion

Reflecting societal changes in other EU countries, the key factor in the transformation of industrial areas in the Slovak Republic is the shift from an industrial to a post-industrial age. However, this gradual process of transformation has been catalysed in Slovakia by other complex transformation processes in the national economy and society, primarily the drastic economic upheavals after 1989 and the current process of European integration.

The revitalisation of distressed industrial areas is, although of some urgency, only marginally treated in the economic policy of the Slovak government; instead the main priority is to provide the optimal conditions for a dynamic economy and increase the country's attractiveness to foreign investors, thus alleviating the most pressing social problem: unemployment.

The impulse for an active policy of revitalisation of industrial areas at national and regional levels is provided by projects financed using EU pre-accession and structural programmes. This has encouraged and necessitated the elaboration of key strategic documents of deliberate regional policy, in which the issues of industrial regions in decline are tackled. These strategies formed the basis for the establishment of Regional Operational Plans at the level of NUTS II and Sectoral Operational Plans in the context of a National Plan of Regional Development of the Slovak Republic. The intentions, objectives and measures were set out in the operational plans at the regional level (NUTS 3), thus creating a basis for the local strategies of regions and districts undergoing structural change.

References

Belcakova I, Gal P (eds) (2000) Current Legislation and Standards of Spatial Planning in Social Transformation and European Integration, SPECTRA FA STU – ROAD, Bratislava

Buček M (1998) Die Ausgangssituation der regionalen sozioökonomischen Disparitäten in der Slowakischen Republik, Paper for the conference of Arbeitsgemeinschaft für den Mittel- und Südosteuropäischen Raum der ARL, Hannover Sušice

Empirica (1993) Produktionsstandorte in West- und Osteuropa, Bonn

Holmanová A, Finka M et al. (1998) Prinzipien, Methoden und Instrumente für eine nachhaltige Entwicklung auf der regionalen Ebene, Zwischenbericht der Forschungsaufgabe VEGA, Bratislava

Finka M (1993) Praktische und Methodologische Aspekte der Integration der SR, Forschungsbericht, Bratislava

Finka M (2000) Background of European Policy in Spatial Planning. In: Finka M, Petrikova D (eds.) Spatial Development and Planning in European Integration, SPECTRA FA STU – ROAD. Bratislava, pp 275-280

Finka M et al. (1997) Implementierung ökologischer Belange in raumplanerische Konzeptionen, ARL/FA STU, Hannover Bratislava

Finka M, Petrikova D (eds) (2000) Spatial Development and Planning in European Integration, SPECTRA FA STU – ROAD, Bratislava

Finka M, Šteis R et al. (1996) Atlas der Raumentwicklung der Slowakei, Methodik der Erarbeitung, I. – IV. Etappe, Forschungsprojekt des Umweltministeriums der SR, Bratislava

Kovac B, Komrska J (eds) (2000) Recent Developments in Urban and Rural Theories and New Trends in Spatial Planning, SPECTRA FA STU – ROAD, Bratislava

Mládek J (1990) Teritorialne priemyslené útvary Slovenska – Territoriale industrielle Gebilde in der Slowakei, Bratislava, UK

Pašiak J (1999) Sídelný vývoj – settlement development, VEDA Bratislava

Rajčák M (1999) The Strategic Intentions and Objectives of the Slovak Regional Policy, Office for Strategy of the Development of the Society, Science and Technology of the Slovak Republic, Bratislava

Rechnitzer J (2000) The Features of the Transition of Hungary's Regional System, Discussion Papers No. 32, Centre for Regional Studies Hungary, Pécs

Steiner M (1997) Competence cluster as a new concept for regional policy, in: Competence Clusters, Workshop report, Graz, Austria

Statistical Yearbooks of Slovakia (1995, 1996, 1997, 1998, 1999), Veda Bratislava

ÚSRSVT SR (1997) The current state of implementation of the Slovak regional policy, Bratislava

Slovenia: Different Types of Old Industrial Regions

Zlata Ploštajner

Urban Planning Institute of the Republic of Slovenia, Ljubljana, Slovenia

1 Historical Overview of Industrial Development

The first stirrings of industrial development in Slovenia can be traced back to the 18th century, when mining, iron, textile, glass and paper industries were built up. The initial ventures were small and determined by the location of raw materials (coal and iron mines, forests). In the 19th century the construction of railroads and the growth of international trade gave new impetus to the process of industrialisation. Industrial centres developed along the railroads (Maribor, Ljubljana, Trbovlje, Celje, etc., see map). However it was in the second half of the 19th century that the process of industrialisation really started in Slovenia.

Political changes during the 20th century greatly influenced the country's industrial development. After World War I and the partition of the Austro-Hungarian Empire, Slovenia became part of the newly established state of Yugoslavia. The foreign economic elite was slowly replaced by a native Slovenian one. As the most developed part of Yugoslavia, Slovenia was able to benefit from the extensive and highly protected Yugoslavian market. The demand for industrial products grew. As a result, Slovenian industrial production shifted from raw materials and intermediate products to final products. During the 1930s Slovenian industry expanded rapidly: the number of industrial enterprises jumped from 275 in 1918 to 523 in 1939. Heavy industries (mining, iron, electricity production), together with the textile and food industries, were in the forefront, with other sectors following.

After the chaos of World War II industrial development soon regained momentum. Accelerated industrialisation was the main component of the socialist economic development policy. The share of industry in the structure of GDP and total workforce grew steadily until 1985. The structure of Slovenian industry diversified further. The primary aim was to satisfy growing domestic demand, final products being produced using imported technology and raw materials. Producers were able to sell most of their products on the highly protected domestic market, despite lower productivity, lower quality, higher production costs and higher prices than foreign producers. Relatively few goods were exported, due to the lower prices they could gain on the world market.

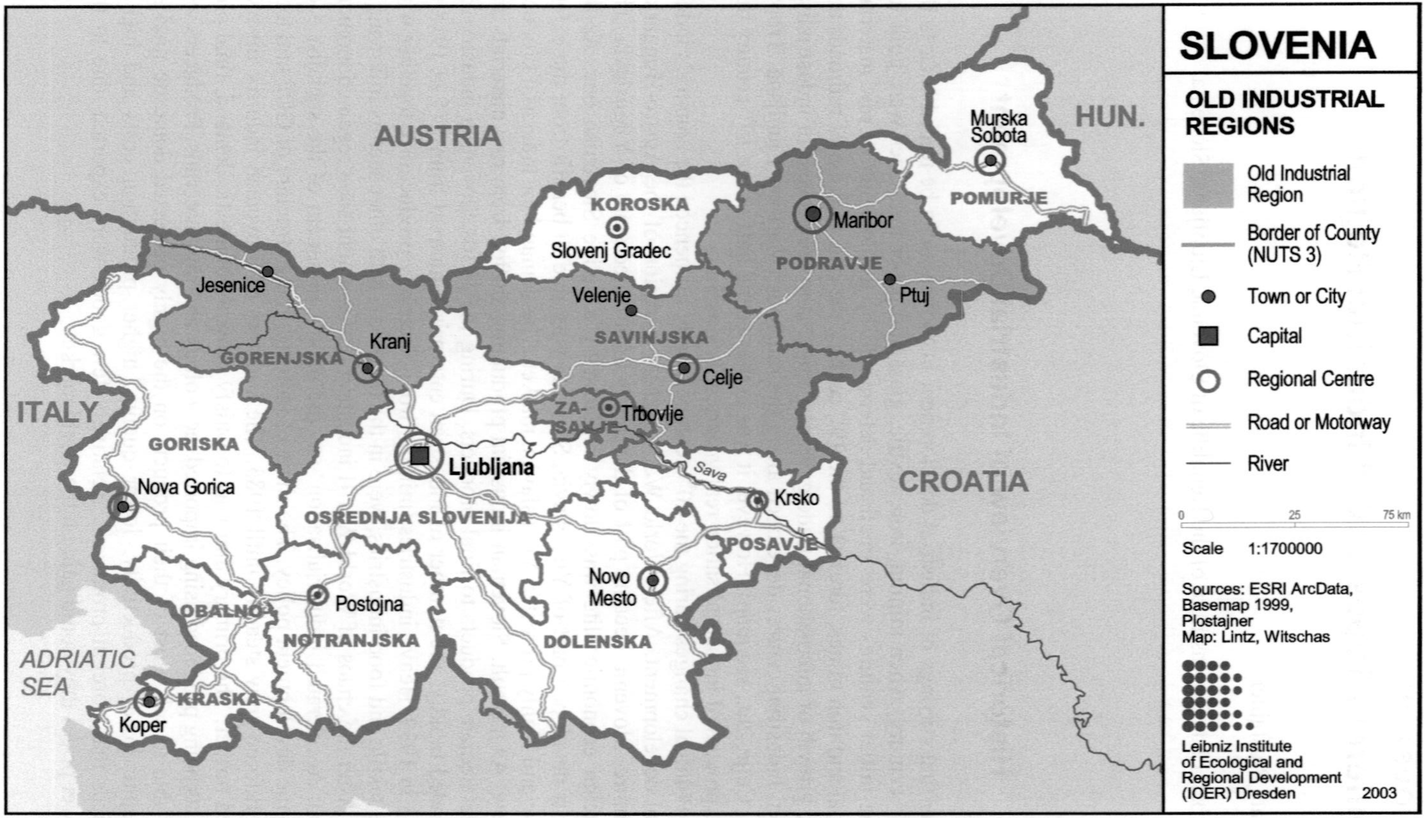

Map: Slovenia – Old industrial regions

The oil crises and a growing external debt put a brake on economic growth. Reforms and measures introduced to deal with the new situation were not successful, and Slovenia missed the opportunity for a timely restructuring of its economy. It is interesting to note that industrial production and industrial employment were declining even before the partition of Yugoslavia and the transition from a socialist to a market economy. As can be seen in Table 1 the start of industrial decline dates from 1986. Employment began to fall, continuing up to and beyond 1997. Production volume, however, started to recover in 1993. The political changes ushered in economic changes. It was anticipated that these changes would prove more painful than if they had been implemented at an earlier date; in the 1990s Slovenian industry faced direct competition from restructured and developed economies.

Table 1. Selected indicators for Slovenian industrial sector

Year	No. of enterprises	Industrial workforce	Production volume indices (1969=100)	Indices of productivity (1969=100)
1970	439	261,065	109	105
1975	426	322,953	163	128
1980	1,920	348.509	213	153
1985	2,084	374,425	238	158
1986	2,109	392,237	242	159
1990	1,460	345,496	211	148
1991	1,490	321,465	185	146
1992	1,630	288,801	160	141
1993	1,605	264,276	156	150
1994	1,657	251,515	166	169
1995	1,613	240,685	169	180
1996	1,530	222,687	171	197
1997	1,480	210,478	173	206

Source: SORS, Statistical yearbooks.

2 Economic Restructuring: Transition from Socialism to a Market Economy

Slovenia has survived the shocks of economic and political transition rather well. However, optimal restructuring has been hampered by the disintegration of the former Yugoslavia, with various negative consequences such as the dissolution of Yugoslavian markets, the loss of industrial property in Serbia, and war in Croatia etc. On a more positive note, Slovenia had the advantage of earlier experience in self-management, based on a quasi-market economy. Furthermore, since the mid-1960s the country had been exposed to democratic and market economic influences, through trade and the free mobility of workers, especially with neighbours Italy and Austria. These factors helped to facilitate the transition in the 1990's.

Now Slovenia can boast a largely stabilised economy, numerous new economic institutions and a legal framework to underpin market functioning. These are important steps on the road to recovery. The formal framework is in place: a major challenge now is to ensure effective coordination and co-operation.

In the economic field emphasis was first placed on macroeconomic stability, with measures to halt the decline in production and employment. As already mentioned in regard to the industrial sector, economic recovery started in 1993. The figures in Table 2 show an initial drastic fall in GDP, with considerable growth afterwards: 1995's GDP exceeded the figure for 1990. Budget deficits and growing debt required special attention at the end of the 1990s. Inflation is still a problem (7.9 percent in 1998 and 6.1 percent in 1999) and the general government consumption of the GDP is around 46 percent.

Table 2. Selected economic indicators for the period 1990-1999

	1990	1991	1992	1993	1994	1995	1996	1997	1998	1999
GDP	-4.7	-8.9	-5.5	2.8	5.3	4.1	3.5	4.6	3.9	3.5
Structure of Value Added										
Agriculture, forestry, fishing (A+B)	5.6	5.8	5.9	5.2	4.6	4.6	4.5	4.3	4.5	4.4
Industry (C+D+E)	37.6	40.8	36.6	34.1	35.4	33.4	32.8	32.5	32.0	31.6
Construction (F)	4.9	4.2	4.3	4.8	4.8	5.1	5.7	5.7	5.5	5.5
Services (G-O)	53.7	51.4	55.0	58.0	57.3	59.2	59.5	59.8	60.3	60.7
Unemployment Rate										
ILO	2.6	7.3	7.4	7.3	7.4	7.4	7.3	7.4	7.9	7.6
Registered	4.7	8.2	14.0	14.4	14.4	13.9	13.9	14.4	14.5	13.9

Source: SORS, Bank of Slovenia, Ministry of finance, IMAD.

Since 1990, a remarkable structural change in market orientation has been achieved: around two-thirds of foreign trade is now with the EU. This has compensated substantially for the loss of the former Yugoslavian market. Today Slovenia's most important trading partners are Germany, Italy, France, Austria and Croatia. Gross fixed capital formation as a percentage of GDP is growing steadily, although this figure is still low compared to countries enjoying high rates of growth. In 1999 the GDP per capita was about 10,078 USD. The share of manufacturing, trade and tourism in the structure of investments is decreasing, while the share of financial, technical and business services is increasing.

One of the most serious long-term problems is unemployment. From 1990-1998 more than 120,000 jobs in industry were lost. This was not compensated by a growth in employment in other sectors. The number of registered unemployed grew from 15,000 in 1987 to 137,000 in 1993; and the total number of employees decreased from 962,000 in 1986 to 742,000 in 1997. Currently the unemployment rate (ILO standardised rate) is still over 7 percent; the registered unemployment rate is about 14 percent (see Table 2). The rate of economic activity of the working age population, formerly among the highest in Europe, dropped from 72 percent in 1987 to 57.7 percent in 1994. The number of employees is around one fourth lower than in 1986, although it has been slowly growing since 1998. Despite the wide range of measures taken in the area of active employment policy, problems in structural unemployment are worsening. This is seen in the high share of unemployed aged over 40, unskilled unemployed (finished only primary school or acquired low level of vocational qualifications) and long-term unemployed. All three phenomena are inter-related; the share of unemployed over 40 and unskilled unemployed increases with the duration of unemployment.

Despite steady economic recovery since 1993, structural changes in the Slovenian economy have meant that industry's share in the structure of GDP has been declining (see Table 2). The future of those industries which cannot compete on the European and world markets is highly insecure. Mining belongs to this group, and most mining ventures (mercury, uranium, lead, coal) have already closed. Many manufacturing firms have also gone bankrupt, while others have undergone a painful process of restructuring.

Today, the structure of Slovenian industry resembles the industrial structure of less developed EU countries, such as Greece and Portugal. The share of low-technology and slowly growing industries is too high. Among the fastest growing and most competitive industries are those producing machinery and equipment, electrical and optical equipment, chemicals (especially pharmaceuticals), rubber and non-metallic mineral products. Many firms are highly export oriented, most of their revenues coming from export to foreign markets. Positive changes in the economic structure are seen clearest when analysing particular branches of industry. Numerous new or improved products and services have appeared on the market, while product ranges have been readily dropped in recovering companies that produce metals, furniture, wood and rubber. This is largely a result of keener competition (foreign trade) and growth in the private sector. Many new small and medium-sized firms have been founded and numerous state-owned enterprises have been subdivided and privatised. The privatisation of state-owned firms in various sectors such as steel, energy, transport, telecommunications and banking has not yet been completed.

Scrapped industrial jobs are only slowly being replaced by new jobs in the service sector (services share in GDP is increasing). Although industry was hit worst by the crisis, the service sector was also affected, especially industry-related services. The growth of services over the last 15 years has been due in no small part to growing demand from industrial firms. It is thus clear that changes in industry have important consequences for the service sector, a fact that should not be forgotten in discussions about a modern services society. Without a strong and healthy industrial sector, further growth in services, especially business services, is questionable.

3 Spatial Patterns

After WW II, the government supported the development of old (e.g. Maribor, Celje, Jesenice) and new industrial centres (e.g. Velenje, Nova Gorica, Koper), with an emphasis on heavy and/or labour-intensive industries. Inherited spatial patterns dictated that northern areas focused on the processing of raw materials, while development in southern areas was more based on new industries (automobiles, electronics, etc.).

In the 1950s and 1960s, industrial development was primarily concentrated in regional centres and in Ljubljana, the capital of Slovenia, leading to a concentration of working families in cities and a depopulation of rural areas. To deal with increasing regional disparities, Slovenia initiated its own regional policy in the early 1970s: in 1971 a special law was adopted on more balanced regional development. This introduced special measures for the promotion of faster development in backward areas. Based on a polycentric concept of development, many settlements were also given their own small branch establishment in the 1970s, providing jobs for local people and halting migration to central cities. This was accompanied by some decentralisation of services from Ljubljana to regional centres, and even to the level of communes (numbering 60 at the time). Regional policy measures in the period 1971-1990 helped reduce disparities between urban and rural areas. However, appropriate measures were still lacking to deal successfully with the decline, since the 1980s, of industrialised regions. Some applied measures have produced unwanted results, such as lower mobility (important for the normal functioning of the labour market). The characteristics of the evolved spatial patterns are as follows:

- Ljubljana and its neighbouring cities developed into the most important economic centre;
- based on inherited patterns of industrial development, the northern part of Slovenia specialised in heavy industry and the processing of raw materials;
- some new industrial centres developed with state support (e.g. Nova Gorica, Koper, Velenje, Novo mesto);
- focusing on a more balanced development, regional policy in the 1970s and 1980s had limited positive effects; it did however contribute to the spread of small labour-intensive industrial branch plants into rural communities;
- based on a polycentric concept of development, commune centres actively participated in industrial and service development, slowly growing into local employment centres;
- owing to the economic crisis and subsequent restructuring, many small industrial plants established in the 1970s and 1980s were shut down, entailing job losses for the local workforce;
- following modern economic trends, restructuring has led to even greater concentration of economic activities (industry and services), especially in Ljubljana and a few other larger regional centres with favourable locations;

- privatisation has affected many aspects of urban life and urban services. Previously many services, both individual and collective (housing, medical services, childcare, leisure provision, and so on) were provided by enterprises. Now the provision of these services is left to local and central government, or to the market;
- daily long-distance commuting became common, with the number of degraded, unused or underused industrial sites increasing.

3.1 Industrial Restructuring at Regional Level

One of the reasons why the effects of transition were not as intense in Slovenia as in other CEE countries is that the country's regional economic base was quite diverse, with marked differences in the structure of regional economies (see map).[1] Two regions could be labelled service regions, although they also have a highly developed industrial sector. These regions are Obalno-kraška, a magnet for tourists, and Osrednja Slovenija (including the capital, Ljubljana). At the beginning of the period of transition and throughout the 1990s they were the most developed regions in Slovenia, with their relative position even improving.

In other regions industry is the prevailing activity. The diversity of regional economic structures has influenced the intensity of adjustment problems (especially unemployment) in individual regions. Some have found themselves in a very tough economic situation. This is especially true of old industrialised regions with primarily traditional industries, such as Zasavje, Podravje, Savinjska and, to a degree, Gorenjska[2]. Dolenjska, by contrast, is a newly industrialised region with a promising economic structure; it has become one of Slovenia's most dynamic regions, attracting a high share of foreign investments. Thus, there exist rather large and, in some cases, growing regional disparities in Slovenia in regard to population, employment, sectoral structure, economic activities and the efficiency of the economy (see Table 3).

Disparities in GDP per capita reflect differences in the regional economic structure and in the restructuring process. Zasavje, Podravje, and Savinjska, formerly belonging to the group of developed regions, are now the most problematic areas; restructuring led to job losses in labour intensive industries in such old industrialised areas, either through bankruptcy or downsizing of production. In the second half of the 1990s, growth continued in the most dynamic regions, i.e. Osrednja Slovenija, Obalno-kraška, and Dolenjska. Some of the previously declining regions, such as Savinjska, Podravje and Goriška, are slowly improving, whilst others are still struggling.

---

[1] Slovenia is divided into 12 statistical regions, also used by Eurostat as NUTS 3 regions. They are used for regional development policy and have no administrative function.

[2] Gorenjska had a much more diversified economy, also in the service sector. Although certain parts of the region were stricken by crises in steel, textiles, basic metal and the leather industry, on average Gorenjska was better off than other old industrial regions. The region benefited from state aid to restructure its ironworks.

Table 3. Regions by size, population and population density

Regions	Area of regions		Population		Density	Unemployment rate in working age population	GDP per capita
	sq. km	% of RS	inhabitants (1998)	% of RS	persons/ km^2	(1998)	(1996)
Osrednja Slovenija	3,546	17.5	517,022	26.0	146	7.3	128.5
Obalno-kraška	1,044	5.2	102,565	5.2	98	7.0	102.3
Gorenjska	2,137	10.5	195,580	9.9	92	8.0	92.4
Goriška	2,325	11.5	119,967	6.0	52	6.0	98.6
Savinjska	2,384	11.8	255,541	12.9	108	11.3	94.5
Dolenjska	1,684	8.3	105,260	5.3	62	7.3	97.9
Pomurje	1,338	6.6	125,441	6.4	94	12.5	77.8
Notranjska	1,456	7.2	50,163	2.5	34	8.2	84.8
Podravje	2,169	10.7	319,617	16.2	147	14.1	82.1
Koroška	1,041	5.1	73,961	3.7	71	8.4	86.3
Posavje	885	4.4	70,187	3.5	79	11.2	90.6
Zasavje	264	1.3	46,633	2.4	177	12.1	84.5
Slovenia	20,273	100.0	1,981,937	100.0	98	7.7	100.0

Source: SORS; Pečar, 1998

Unemployment is a continuing problem in many regions (see Table 3). In dynamic regions such as Dolenjska, Osrednja Slovenija and Obalno-kraška, unemployment is decreasing due to economic growth; in some regions unemployment is decreasing only due to the commuting of workers to other regions (Gorenjska) or migration. In other stricken regions the unemployment rate is still very high.

The poor educational profile of industrial regions undergoing structural change will severely limit future prospects for economic improvement unless action is taken in the next few years. Better standards in education should lead to increased labour mobility, promotion of entrepreneurship and innovation, as well as the development of economic awareness and acceptance of a more market-oriented society. The mobility of labour in Slovenia is much lower than in other developed market economies. This can be attributed to several factors: an insufficient adaptation of qualifications to new skill requirements in different occupations; a reluctance to change jobs, ingrained in the workforce from the permanent character of employment in the previous socialist type of economy; passive attitudes of employees and expectations of state assistance; finally, the housing barrier should also be mentioned, as this restricts labour migration within the country. Training schemes aimed at retraining or upgrading job skills, as well as providing information on self-employment and the founding of new enterprises, should be decentralised and based on local initiatives.

Restructuring has led to a growing spatial concentration of jobs. Labour migration as an important spatial phenomenon in Slovenia clearly shows the changes. The municipality of Ljubljana alone provides more than one fifth of all jobs in Slovenia, attracting workers from all over the region and neighbouring regions (especially Gorenjska). Its gravitation area covers 30 percent of the territory of Slovenia and 38 percent of the total population. Polarisation between Ljubljana and other parts of Slovenia has increased. Being the capital, Ljubljana has also benefited from the expansion in services. Labour concentration is also characteristic of other regions, with data showing that eleven urban municipalities together have more then half of all jobs. Maribor (Podravje), Celje (Savinjska), Novo mesto (Dolenjska) and Murska Sobota (Pomurje) are regional employment centres, while all others are only local centres. This situation has forced many people to move closer to their jobs, reflected in population changes during the last decade. The harsh economic situation has led to a decreasing population of Savinjska, Podravje and Zasavje. In contrast, Dolenjska's population is growing. The same is true for the most dynamic regions, Osrednja Slovenija and Obalno-kraška.

Below, the problems and strategies for two different types of old industrial regions are presented: firstly, a monostructured region – Zasavje – and then two regions with more diversified but outdated economic structure – Savinjska and Podravje.

Zasavje

The most highly monostructured region is Zasavje, this also being the smallest region (population: 46,800) with the highest population density (177/km^2). Previously it was one of the most developed industrial regions, with an economy based on mining and electricity production. Other industrial activities in Zasavje were also more or less directly connected with these industries. This monostructured economy meant that the region was very sensitive to economic changes. Stagnation gradually set in during the 1980s, and some unsuccessful measures were undertaken to change the course of developments. In 1991, two-thirds of all jobs were still in industry, considerably above the national average. The situation in mining and electricity production worsened throughout the 1990s with the loss of many jobs; the registered unemployment rate is now about 20 percent. Many workers have left the region, while others commute daily to the more favourable job market in Ljubljana. This migration of labour explains the big differences in gross value added and income tax base per inhabitant (incomes are earned outside the region).

When the national government decided to close most of Slovenia's mines, the region's future became even more uncertain. Coal from the Trbovlje-Hrastnik mine is mainly used by steam power plants, the biggest located in Trbovlje. One solution proposed was the building of an additional plant in Trbovlje; however this has been abandoned under growing ecological standards and for economic reasons (liberalisation of the energy market). Even the long-term future of the old plant is in doubt.

In an attempt to resolve the difficult situation, the region has started to organise and lobby for itself. A highly active regional development agency was established early on. This agency prepared a successful development programme for the region, and restructuring of the region has become a national issue. The law on the closure of the Trbovlje-Hrastnik mine (Law on closure of the Trbovlje-Hrastnik mine and development restructuring of the region, 2000) has, together with financial resources, provided jobs connected to the closure of the mine (around 70 million euros for technical activities, ecological rehabilitation of sites, re-employment of miners, etc.). Furthermore, additional funds were provided for the economic restructuring of Zasavje in the period 2000-2008. In the five years from 2000-2004 the region will receive around 20 million euros to build an economic development infrastructure, for human resources development, and for the promotion of investments and economic restructuring. The amount for the next four years has yet to be specified.

One of the most important goals is to support both the restructuring of already existing enterprises and the development of new ones. Although the level of entrepreneurship in Zasavje is far below the Slovenian average, manufacturing firms do exist which could be successfully developed. They are the main contributors to regional exports (90 percent of all export revenues). The most important industries are the production of electrical and optical equipment, non-metallic minerals and chemical products. However, these industries need support to upgrade their production technology and to improve marketing and management. They want to establish a common technological and innovation centre, enabling them to pool their research and development resources and achieve synergies. A training centre for manufacturing will be established in the field of human resources and regional education. In addition, a labour fund should be established to offer special help to workers.

Zagorje ob Savi: A Mining Town Makes Its Way

The municipality of Zagorje ob Savi (pop. 17,000) is situated in the Zasavje region in Central Slovenia. Once a mainly forested area with wide valleys and few settlements, the area's appearance has changed dramatically since the beginning of the 19th century. A railroad system, constructed from 1846 to 1849, accelerated mining activities in Zasavje. Two textile plants and one electro-technical plant were established to provide employment for women. These developments led to population growth in an area not well suited for settlement. Open-cast mines and quarries devastated the countryside, while redirecting waterways and affecting local vegetation; underground excavation caused landslides and ugly dumps of waste material. Now the land is being repaired, e.g. by forestation in order to prevent slides. There is an ongoing process of pit filling, and a system of ditches is being constructed on the levelled ground to allow water drainage. Advanced techniques are being used to minimize depressions, land slides etc. Mine tunnels and pits are to be completely filled in the next years.

Mining is no longer the prevailing industrial branch in Zagorje ob Savi, and a project to close the mine is underway. Electro-technical, wood, building and textile industries are now firmly established and are more important for the local economy. After concentrating on mining for so many years, the town has now transformed itself from a grey, dirty mining area to a pleasant green town. The former dependence on mining could only be solved by focusing on small workshops, producing a variety of products. Zagorje ob Savi is on the right path, with several successful artisans emerging in the early nineties. However Zasavje region still suffers from high unemployment, a low percentage of high educated employees in the secondary and tertiary sector and a high rate of migrant workers.

The economic crisis stimulated the municipality to establish a special Development Agency. One project to establish an industrial and small business zone on former production facilities near the town was successfully launched in 2000. The zone for artisans and enterprises is supposed to employ about 300 people. The European Union will assist in the establishment and initial activities of an innovation and development centre, as well as a centre for international co-operation. A contract with the Phare programme has already been signed. The municipality Zagorje ob Savi would benefit from development of its tourist potential. As a first step some mines could be opened for visitors to give insight into the miners' way of life. One museum has already been opened and the opening of Markov Stangovc, a carsic cave discovered in 1968, is planned in the next few years. Another important issue is that of raising awareness of environmental problems. For example, one elementary school is currently participating in an eco-schools project.

Manca Prašnikar, Community of Zagorje ob Savi

Savinjska and Podravje

Savinjska and Podravje have a long tradition in industrial production, particularly heavy industry and the manufacture of intermediate and final products. Demand for these products fell after the dissolution of Yugoslavia and as a result of tight restrictions on access to the EU market, leading to overcapacities in many branches. Important spatial differences can now be observed between the two regions in their ongoing process of restructuring.

Savinjska is an old industrial region with a quite diversified economic base, ranging from heavy industry (ironworks, power plants, mining) to consumer goods, from tourism to business services. The region has two development poles – Celje (traditional industrial centre) and Velenje (new industrial centre) – and the economic recession which started in 1989 and developed into crisis by 1992 affected the two centres differently. Previously, economic development was associated with big companies. In Velenje most of these have survived by adapting to the changed conditions and quickly establishing new development cycles. This contrasts to the situation in Celje. Owing to the disintegration and bankruptcy of many big companies, unemployment in Celje has been very high since 1992. Around 10,000 jobs were lost in only a few years, as labour-intensive companies were hit severely by the recession. The crisis accelerated in 1995 and 1996, with recovery only beginning slowly the following year. Despite the important changes, industry is still a leading sector. In 1997, industry contributed 66 percent of the gross value added and provided two-thirds of all jobs. Today the most important industries are the manufacture of machinery and equipment, food, tobacco, textiles, chemicals (especially pharmaceuticals) and non-metallic mineral products. Although some companies are competing successfully on the world market, the overall economic situation of the region is dissatisfactory.

Savinjska has the advantage of a good location. Its accessibility will increase further with the construction of new motorways, making the area more attractive for investment. The regional agency established in 1996 has been actively seeking foreign partners to invest in Savinjska. As a result, some big companies have been able to survive and consolidate, functioning as the generators of future development. With good possibilities for the development of tourism and services and with slowly recovering industrial sectors, the future looks promising. The most important regional project is the establishment of a Slovenian centre for heavy equipment (with state co-financing), to serve as a nexus for an emerging regional cluster.

The transition in Podravje region, with the traditional and thriving industrial centre of Maribor being second in importance only to Ljubljana, has had many negative aspects. Disintegration of big companies, large overcapacities in metal, automobile and machinery industries, outdated programmes and products, and the lack of a high quality managerial and professional cadre have been the main reasons for the dire situation of Podravje's economy in comparison to the rest of Slovenia. The state did not improve matters with inappropriate actions and general slowness to react. Following bankruptcies and workforce cuts, Podravje has suffered the highest unemployment in Slovenia. The slow process of recovery started in 1997. Although Podravje's industry was badly hit, it still produces almost half of the region's gross value added. Podravje is actively seeking foreign investments, and some companies have managed to find foreign partners or foreign buyers.

3.2 Degraded Urban Areas

The restructuring process has brought many spatial problems to old industrial cities, leaving them with weak local economies, few sources of revenue, many social problems and vast areas of derelict and abandoned land as a result of industrial decline or the relocation of production to other sites. These areas are very often ecologically devastated, and special treatments and resources are needed to enable reuse. In view of their structure, size and location, degraded urban areas are valuable assets for the redevelopment of cities. They can be redeveloped for diverse usage, sometimes offering new opportunities for the restructuring of city centres (e.g. Maribor, Jesenice, Kočevje), for rerouting traffic links, and for resolving other specific functional problems (Koželj 1998).

Industrial areas, mining areas and transitional areas comprise three-quarters of all degraded areas in Slovenia. More than one half are industrial areas. It is expected that mining areas such as in the Zasavje region will be renaturalised (open countryside, or restored to agricultural use). Industrial areas are compact complexes, often covering extensive areas. Some have already been transformed for other uses (commercial, services, modern business zones, etc.). Local governments, with no effective land policy and planning instruments at their disposal, can not direct and regulate this transformation. Division and subdivision of uniform and compact areas has very often resulted in problems of accessibility, incompatibility of usage, poor maintenance of infrastructure, etc. Accessibility needs to be improved by the introduction of public spaces and new access routes, although this is hard to realise owing to private ownership of land. In old industrial cities, large rundown housing estates can also be viewed as degraded areas offering possibilities for regeneration.

4 Regional Policy and Orientations

Slovenia was not well prepared at the regional level to deal with the problems of transition (Ministry of Economic Relations and Development, 1999). In the 1990s, regional policy was oriented towards rural, depopulated, and less-developed areas. These areas were hit by the transition and restructuring, with many small plants being forced to close. However it must be said that the decline of old industrial centres has had much more serious implications. The state and individual cities lack the resources to deal with the range of problems facing the country, problems with technological, ecological, regional, economic, social and political dimensions. In the period of transition the problems of industrial restructuring are myriad and acute, with a similar combination of factors as those undermining the economy as a whole. Since these problems have been wholly ignored by Slovenian regional policy, the government has been treating them in most cases as sectoral, rather than regional, problems. In the absence of a comprehensive regional industrial policy, the Slovenian government has been helping troubled firms on an individual basis, with no regional framework or predetermined criteria.

In 1999, a new regional development law was adopted, providing a framework, as well as instruments and measures, to deal with declining regions. Following EU practice, three priority areas were defined, corresponding to Objective 1 areas (peripheral agricultural and mountainous, population decreasing), Objective 2 areas (with old industry and high percentage of unemployment) and border areas (within 10 km of the border). The aim of the law is to provide coordination between regional policy and all sectoral policies with regard to principles, targets, strategies and instruments that need to be established for effective regional development policy. The task of coordination is the responsibility, at the national level, of the Council for Regional Development and the National Agency for Regional Development and, at the regional level, of regional development agencies and regional development councils.

A key goal of Slovenia's economic development strategy is achieving sustainable economic growth, thereby catching up with the more developed European countries; increased competitiveness of the Slovenian economy is the means to realising the goal. This strategy helped shape the National Development Programme for the period 2000-2002. Two of the worst affected old industrial regions, Zasavje and Savinjska, were given special treatment by the programme, with Prekmurje region also being designated for special help. The respective regional development programmes were incorporated into the national programme, with the state agreeing to contribute considerable additional financial resources in accordance with EU rules.

Changes at regional level over the last decade have shown that market forces alone cannot ensure balanced development. Economic development as directed by the market has been concentrated in already developed regions. Disparities between the least and the most developed regions are now at a ratio of 1:3 in terms of value added.

Demand for available locations is growing with business expansion. Unfortunately, complicated and cumbersome planning procedures mean that municipalities cannot respond quickly to this demand. The supply of larger plots is a particular problem, and municipalities are therefore attempting to establish and develop business zones. Most of these can, however, only satisfy the needs of SMEs, unable to finance the purchase of land for future business development. Municipalities rely upon the land they already possess and on the readiness of private owners to co-operate. They lack an effective land policy and land planning instruments with which to direct and regulate the development processes in their communities. One particular problem is the absence of administrative regions and regional planning in Slovenia. Future socioeconomic development patterns should be defined at this level, since most municipalities are small (there are 192 municipalities in total) and have very limited capacities at their disposal, both human and financial. The importance of regional institutional building was overlooked until recently.

For that reason, regional and subregional development agencies have only now been established to coordinate all development activities in their area. With their new professional staff, these agencies will assist municipalities in fulfilling their development goals. The future success of Slovenia's cities and regions will depend on their joint efforts to foster an economic and political climate that allows businesses and industries to compete successfully in domestic and international markets. To this end, a dedicated regional institution must be given the responsibility of coordinating spatial planning activities at the regional level.

References

Bank of Slovenia: Annual reports for relevant years

Dimitrovska Andrews K. et al. (1997) Zasnova urbanizacije (Urbanisation Trends in Slovenia), Faza 7, Urbanistični inštitut Republike Slovenije (Urban Planning Institute of the Republic of Slovenia), Ljubljana

Dimitrovska Andrews K, Ploštajner Z (2000) Local effects of transformation processes in Slovenia, Informationen zur Raumentwicklung. No. 7/8, pp 435-449

IMAD: Economic mirrors for relevant years

Koželj J (1998) Degraded urban areas, Ljubljana: MOP

Kukar S, Gulič A (1992) Regionalni razvoj in regionalizacija Slovenije, Ljubljana: Inštitut za ekonomska raziskovanja and Urbanistični inštitut Republike Slovenije

Law on closure of the Trbovlje-Hrastnik mine and development restructuring of the region (2000) In: Official Gazette RS, No. 61, Ljubljana, pp 7675- 7676

Law on promotion of balanced regional development (1999). In: Official Gazette, No. 60, Ljubljana, pp 7674-7678

National Development Program for the period 2000-2002, Republic of Slovenia, Ministry of Economic Relations and Development

National Employment Office: Regular reports on unemployment

Pečar J (2000) Regionalni vidiki razvoja Slovenije s poudarkom na finančnih rezultatih poslovanja gospodarskih družb v letu 1998. Ljubljana: Urad Republike Slovenije za makroekonomske analize in razvoj

Statistical Office of the Republic of Slovenia: Statistical yearbooks for relevant years.

White Paper on Regional Development Strategy, draft, (1999) Republic of Slovenia, Ministry of Economic Relations and Development

Ukraine: Transformation of Economic Structure as the Key to Spatial Development

Leonid Rudenko, Irina Gukalova and Gregory Pidgrushniy

National Academy of Sciences, Institute of Geography, Kiev

1 Introduction

Ukraine's regions must be considered individually when analysing the country-wide transformation under the new economic conditions. Each region shows both positive and negative features, and these must be clearly evaluated according to the changing situation. Thus one important step in formulating spatially differentiated solutions to the economic, ecological and social problems facing the country is to analyse the different characteristics of the regions and to define spatial categories. Industrialised regions are seen to be particularly weak, and according to some economic indicators they have developed more slowly than balanced industrial-agrarian or agrarian regions. At the same time not all of these regions can be called regions in decline. In view of their long historical development, all industrial regions in the country can be designated "old", with some structural adjustment to the new economy being necessary. The economic diversity in most industrialised cities and regions is a positive factor in overcoming the crisis. On the other hand, a crisis in any well developed mono-structured industry can lead to the economic destruction of the whole region. Highly developed and diversified regions have inherent advantages which enable them to take a lead in positive restructuring. A study of the relevant factors of restructuring and their interrelations constitutes a second step for a positive development of the Ukraine.

2 General Socio-Economic Situation in the Ukraine

2.1 Geographical Situation

Situated in eastern Europe, at the north coast of the Black Sea, the Ukraine is a sovereign state, having gained independence from the former Soviet Union in 1991. It shares borders with Russia and Belarus in the north and north-west, and with Poland, Slovakia, Hungary and Moldova in the west.

Map: Ukraine – Typology of oblasts

The total area is 603,700 km². For administrative purposes, the Ukraine is divided into 24 oblasts and one autonomous republic (the Crimea, see map); the capital city is Kyiv. Most of Ukraine's territory (95 %) is lowland. The only mountainous areas are the Carpathians in the far west and the Crimean mountains in the south. Nearly three-quarters of the land is forest steppe and steppe zones. Two natural resources – soils and minerals – have formed the basis of the development of agriculture and heavy industry, and both sectors are significant in the economic structure of the country.

Today, almost all of the country's territory is used for economic production. The share of technologically transformed land, including built-up areas, open pits, open cast mines and other mining areas, is larger than in most other European countries (Gorlenko and Rudenko 1994). The population density is also above the European average. Consequently, relatively little land can be said to be in a "natural" state. For example, forest covers less than 15 % of the country – half the world and European average. Natural resources have always been highly important to industrial development, especially coal, iron and mineral ores. Extraction, enrichment and processing activities have determined the high concentration of mining industries in several Ukrainian regions.

A good example is the Donbas region (Donetsk and Luhansk oblasts), one of the largest old industrialised regions in Europe, with a high concentration of mining and heavy industry. Second in size and importance is the industrial region Near-Dnieper (Zaporizzhya, Dnipropetrivsk regions). Both regions are situated in the east of the country. Other industrial regions in the south (Odessa, Mykolayiv, Kherson) and west (Near-Carpathians) or in big cities, are generally characterised by the late formation of various industrial structures, associated with the development of large numbers of high-tech branches and industries. Here the city of Kyiv holds a special position. Despite being an old industrial centre, it has benefited from a high level of investment in high-tech industries and non-productive spheres, leading to changes in its economic structure.

2.2 Development in the Past

At the beginning of the 20th century the Ukraine had about 5,000 industrial enterprises, employing around 300,000 workers. By 1914 the industrial workforce had risen to 630,000. Growth rates were high and production was concentrated (Ukraine produced 65 % of the Russian Empire's coal output and 51 % of its iron output), but the overall structure of industry was deformed, with both sectoral and geographical disparities. Thus mining was responsible for 60 % of total production, compared to 10 % from machine-building and metal-processing branches.

The location of industrial enterprises was highly irregular throughout the Ukraine, being concentrated mainly in the old industrial region of Donbas, the Near-Dnieper area, and major industrial centres such as Kyiv, Kharkiv, Odessa, and Mykolayiv. The rest of the Ukrainian territory, with the exception of some cities (usually the administrative centres of provinces), remained industrially underdeveloped. The early years of socialism, beginning in 1917, were characterised by complex processes associated with the establishment of a centrally planned economy, the rebuilding of the country after the civil war of the 1920s, and subsequent industrialisation. Industry was developed to serve the expanding military sector, with emphasis placed on heavy industry. In 1940 industrial output was more than seven times the 1913 level. Despite even higher rates of growth in machine-building (1940 output 50 times the 1913 figure), electricity generation and other industries, the structure of industrial production remained deformed. The consumer goods sector, the chemicals and electro-technical industries, and the production of electrical equipment and heavy vehicles, were all lagging behind. During World War II the Ukrainian economy suffered heavy losses, with more than 16,000 enterprises being destroyed.

In the next stage of industrial development (from 1945 to 1965) new industrial assets were constructed, with older assets being re-built and renovated. However, even with high rates of industrial growth, these processes did not translate into a wider usage of improved technology or a diversification of production. By the late 1960s Ukraine's industry had been redeveloped. In the 1970s priority was given to machine construction, the aviation industry, automated machine-building, electrical equipment, electronics, organic chemicals, etc. Western Ukrainian regions (Trans-Carpathian, Lviv, Volyn, Rivne, Ternopil', Ivano-Frankivsk, Chernivtsy) and central regions (Chmelnytsky, Vynnytsya, Cherkasy, Kyiv) benefited from the newly developing industries.

Industrial development in these regions was favoured by the existence of a well-developed production base, good geographic conditions and a sizeable workforce. The regions in the south and east of the Ukraine (Poltava, Kharkiv, Dniepropetrovsk, Donetsk, Kyrovohrad, Mykolayiv, etc.) saw the least development of industrial structure. Innovatory patterns were adopted in Kharkiv and Dniepropetrovsk, regions, for example, where conditions of industrial and territorial division (from the modernisation and partial re-specialisation of existing industries) proved favourable. Other older industrial regions, e.g. Donetsk, where the industrial infrastructure had been formed at a much earlier stage (with coal industry, ferrous metallurgy, and low-technology branches of heavy machine-building), proved too conservative to incorporate technological innovations. In most cases new technologies were accompanied by a development of priority industries and an expansion or formation of new industrial centres.

In the 1980s the Ukrainian industrial complex played a very important role in the Soviet economy. In terms of the volume of industrial and agricultural output, Ukraine was second only to the Russian Federation. The Ukraine was responsible for around 20 % of the USSR's GDP; its share of Soviet industry and agriculture were 19 % and 24 % respectively. These figures are all the more impressive when one considers that geographically the Ukraine made up only 2.7 % of the USSR land area; however its population was 18.2 % of the total. With its active participation in COMECON, the Ukraine was important to many Soviet industries. It produced 20 % of phosphorous fertilizers, 30 % of coal, 34.5 % of steel, 35 % of rolled metal, 41 % of iron, 72 % of manganese ore, 72 % of nitrate fertilizers, 97 % of engines and 100 % of corn and beetroot harvesting machines. Despite the necessary preconditions, Ukrainian industry was not transformed to a higher level in the 1970s and 1980s, as was the case in developed Western nations. Thus the chance of realising a post-industrial economy was missed (Pidgrushniy 2000).

On the contrary, this period was characterised by a growing stagnation of production. The crisis began in the second half of the 1970s and the first years of the 1980s, when rates of industrial growth began to fall rapidly. Following the breakdown of the USSR in 1991, the processes of transition to a market economy and integration into the global economic system began; reforms, however, were slow and inconsistent, tending to exacerbate the overall economic crisis and an economic collapse was the end result. In 1990 industrial output stagnated, and the economy subsequently contracted. In 1999 things began to look up and growth was finally resumed (+ 4 %).

The reasons for the chronic problems afflicting the Soviet economy were rooted in the genesis of the country's industry, the peculiarities of its planned industrial evolution, and the subsequent attempts to introduce market economies. All these factors demanded clear thinking by the ruling elites, with the adoption of adequate policies and the necessary organisation measures to ensure structural changes in government and, most importantly, in industry. At the end of 1980s the communist leadership adopted a state programme of "Acceleration on the basis of machine-building"; this did not produce any positive results. Moreover, attempts at price liberalisation to help promote foreign economic and entrepreneurial activities only sped the collapse of the socioeconomic system. By ignoring economic value laws, structural adjustment was blocked; this gave rise to such phenomena as "the system's inability to respond to scientific and technical progress" (Doroguntsov and Gorska 1998).

There are additional reasons, besides the usual drawbacks of a command economy, to explain the slow modernisation of production and unimpressive productivity. For example high technologies remained entirely in the domain of the military sector, and were not adopted to help modernise the civil economy. Both the consumer sphere and the service sector were underdeveloped (the latter stimulates investment into innovative research in a market economy).The Ukraine inherited 35 % of the former military sector of the USSR; in a deformed branch structure heavy industry was massively over-represented (70 %). All this prevented progressive restructuring and hindered a speedy transition to a higher stage of industrial development.

2.3 Transformation Processes in the Ukraine

The ongoing processes of transformation in the former Soviet republics take a variety of forms. The Ukraine holds an unenviable position among them: initially regarded as one of the most developed post-Soviet countries, it sank in 1992 into an abyss of economic depression, from which it has not yet been able to free itself.

In early 1990s the Ukraine, the second largest country in eastern Europe (the population exceeds 50 million), had, according to all western experts, an extraordinary potential for advancement. This potential was especially visible in agriculture, with a unique combination of favourable climate, highly productive soils and a proximity to international markets. Ukraine's industry also had a great capacity for growth. Supported by a highly-skilled labour force, the country was expected to do well in an international competitive environment, with steady progress towards a successful integration into the global economy. However, few positive economic developments were recorded in the first decade of independence. In 2000 GDP recorded its first positive growth (6 %) since the breakdown of the USSR, with industry growing by 12.4 %. This growth was in part determined by transient factors; however the privatisation of market-oriented industries from previous years also played a role.

Any process of modernisation of production and change in social life presupposes the existence of an active social subject. At the stage of advanced industrialised development, technocratic elites constitute such a group (managers of industrial enterprises, specialists, government agencies, etc.). After Ukraine gained independence the former Soviet technocratic elite retained control not only over industry, but also maintained real political power in the country. However these leaders were unable to move the country to a post-industrialised condition, failing either to adapt the production system to current market conditions or to create a favourable legislative and economic environment. This explains the lack of economic investment. The necessary social forces which have previously led to a post-industrial transition in developed countries, i.e. a new generation of managers and professionals, or a powerful green movement (Khristov 1996), have not yet emerged. Improved conditions and more time are needed to realise these forces.

Despite the reforms of the last years, the state apparatus does not have a rational structure for balanced regional development. Instead, a complex structure of ministries and agencies exist (more than 100 organisations), the majority of which focus on particular sectors. The country lacks a strong centre for the planning and coordination of economic and industrial policy, although the Ministry of Economy provides some leadership in this regard. It duplicates many functions of other sector-oriented governmental ministries and agencies and depends heavily on decisions of the Cabinet of Ministers. This structure is more or less copied at all regional levels. The overall impact on transformation processes is negative, as all attempts at external and internal liberalisation and de-monopolisation of Ukraine's economy are impeded. It should also be borne in mind that Ukraine's economy is highly dependent on world and regional markets (Russia, the EU, east and central European countries, and the US) as well as on internal political stability. Any unexpected disruptions to these or other spheres can influence rates of economic growth and regional economic policy. Although a high dependence on world markets does not normally pose a serious problem, in the case of Ukraine's predominantly mono-structural export orientation (with primarily metallurgical products) it constitutes a considerable weakness. Large subsidies in the form of tax breaks or reduced energy costs are necessary to maintain competitiveness.

The government adopted a long-term programme in March 2000, determining a complex and thorough strategy of economic reform. Unfortunately the speed of realisation is very low. Only modest progress has been achieved in the privatisation of energy and other sectors and in the reform of infrastructure (communications, transport).

Important for the economic development of old industrial regions are Special Economic Zones (SEZs) and territories of priority development. The development of SEZs in Ukraine's legislation has taken about 10 years. 1991's Law on "Foreign economic activity" launched a series of legislative acts in this sphere. The status and size of these zones are determined by the Ukrainian Parliament, Verkhovna Rada, by adopting a special law on each zone. The next important step in this direction was the passing of the Law on "General Principles of Creation and Functioning of Special (Free) Economic Zones" (1992). This important law was adopted without changes or amendments, despite expert opinion that it may prove insufficient. The law determines goals in the creation of an SEZ, such as attracting foreign investments, enhancing the investment climate, fostering entrepreneurial activities with foreign investors, increasing the export of goods and services, the introduction of new technologies and the supply of high quality products and services to the internal market. Further aims are to improve market methods, the market infrastructure and the use of natural and labour resources. In terms of concrete results, SEZs have still to justify their existence. Thus, according to data from the Cabinet of Ministers, the amount of investment attracted to SEZs is 305 million US Dollars – 24 % of the total amount budgeted for approved projects. This money was intended to create 25,000 new jobs and secure the future of 37,000 current jobs. However, only 30,000 jobs were in fact preserved. In total more than 100 laws and legislative acts have been issued in connection with special economic zones. There are now 11 SEZs in 9 regions with special investment regimes, e.g. in the Donetsk region, Luhansk region, Chernihiv region and in the towns Shostka and Kharkiv.

In the summer of 1999 another package of economic statutes for the creation of SEZs and the introduction of a special system of investment measures was prepared for territories of priority development, primarily regions of decline. The legislation aimed at establishing a positive investment climate, securing jobs, stimulating business activity and shifting the focus of socioeconomic processes to the regional level, with increasing regional responsibility for results. Preliminary estimates from the Ministry of Economy indicate that the introduction of SEZs and the special measures in support of investment activity in territories of priority development will create 200,000 jobs over the next 5-7 years. The following factors are critical in the promotion of internal production (including the shifting of "grey" industries into the formal economy) and for attracting foreign investments: a liberal internal and external economic regime, with macroeconomic and political stability; transparency and relative stability of legislation, tax and customs legislation; simplified administrative procedures, the possibility of effective court redress, clearly defined and protected ownership rights, an absence of administrative pressures; a functioning transport and communications infrastructure. Special tax breaks and the creation of free economic zones constitute only the first step in the right direction.

2.4 Problems of European Integration

The Partnership and Co-operation Agreement (PCA) which came into force on March 1, 1998, defines the fundamental principles of the political and economic relationship between the Ukraine and the European Union (EU) while further strengthening co-operation based on the common values of both sides. The PCA allows the co-ordination and consolidation of political and economic strategies, and from an economic perspective it constitutes an important step towards Ukraine's accession into the legal structures of a common European market and the World Trade Organisation (WTO); it is also a tool to foster the development of a market economy in the Ukraine. The full implementation of the PCA and the ultimate realisation of the country's potential is the highest priority in the Ukraine-EU relationship. The PCA leaves room for manoeuvre in a wide range of issues, so that the relationship between the two partners can be further consolidated and elaborated.

Indeed, the extent and dynamism of further rapprochement between the Ukraine and the EU in different fields has been left undecided, ultimately depending on the willingness of both parties. At the European Summit in Helsinki in December 1999 the EU Council approved the EU Common Strategy towards the Ukraine, supporting Ukraine's integration into the European economy and consolidating its European identity. According to a decree issued by Ukraine's President in September 2000, the programme for integration into the EU has already been prepared, with the programme structure taking account of the experience of other EU candidate countries. When the Ukrainian authorities closed down the Chernobyl atomic power station in December 2000, an important symbolic step was taken in furthering ties with the European Union. Political events at the beginning of 2001 incited the Parliamentary Assembly of the Council of Europe to consider barring the Ukraine from membership for human rights violations. This shows how internal political instability can quickly lead to distrust and protest from other European states, and reiterates the need for the Ukraine to continuing in its efforts towards a civilised accession into the European community.

3 Characteristics of Specific Problems Faced by Old Industrialised Regions

3.1 Economic Situation

Today the national economy is in the throes of an ongoing crisis, caused and sustained by both primary and secondary factors. The former include a decline in production volumes, a degrading of the economic structure, and declining markets (mainly in the processing industry, with shrinking markets for high-tech products). The latter secondary factors encompass financial instability, sluggish investment processes, decreased tax revenues, growing domestic and foreign government debt. GDP serves as a decisive evaluation indicator in modelling current economic transformations – in 1990 it was 207.7 billion Ukrainian Hryvnia (UAH, 1996 prices level) or 111.6 billion US$. By 1998 it had fallen drastically to 101.1 billion UAH or 41.3 billion US$. A good GDP growth of 6 % and an even better growth of industrial production of 12.9 % were finally seen in 2000, and some growth of earnings was also observed. However, the overall industrial productive capacity of the regions had fallen, with three regions producing 47.3 % of all industrial products – Donetsk, Dniepropetrovsk and Zaporizzhya regions. Their share in Ukraine's export stands at a similar level. Half of Ukraine's regions (with the lowest export rates) produce less than 10 % of exports. There are considerable differences between Ukrainian regions and links between them are growing weaker; the highest level of inter-regional economic ties only exists among Donetsk, Dniepropetrovsk, Luhansk and Zaporizzhya regions – all old industrialised regions.

The following economic developments can be regarded as benefiting the market: de-nationalisation of property; the emergence of a non-governmental economic sector and private enterprise; budding national goods, labour and capital markets, with prices being set by the market. The country is developing a national financial and banking system as well as a regulated currency market. Furthermore, a diversification of foreign economic relations and an identification of new channels of commercial and investment collaboration is being realised. Another positive factor is the willingness displayed by the government, by the business community and by the public to endure unprecedented economic constraints, such as hyperinflation, regular non-payments, lack of social protection, etc. The nation has presumably developed a certain immunity against spontaneous actions taken by economy management bodies.

Negative phenomena and trends are Ukraine's concern. Thus the public administration structure has not been modernised according to the objective laws of a market economy. In particular, the processes of privatisation did not lead to efficient ownership; a viable system of market institutions has not been adopted, high-tech production facilities have been lost, as well as former leadership in numerous areas of fundamental research. Brain-drain has reached critical proportions. The real and financial sectors of the economy are highly unbalanced. Domestic and foreign government debt is growing fast, with the likelihood of defaulting on servicing obligations. The domestic investment crisis has become systemic, and direct foreign investments are evidently insufficient. On October 1, 1999, the total amount of direct foreign investment in Ukraine was 2.9 billion US$, a figure pitifully small in comparison with other eastern European countries. Rates of investment flow have been diminishing drastically since 1994.

The investment climate in the Ukraine has been continuously deteriorating. A series of government decrees aimed at attracting foreign investment has had no positive effect. Investors have re-orientated on enterprises with a fast return of capital and maximum profits. This explains the interest in the retail sales sphere (15.8 % of investment) and food industry (21 % of investment). External loans from international financial organisations, indispensable for an economy in transition, have in fact had negative effects by depressing initiatives aimed at identifying alternative domestic funding for social needs. The population as a whole has become poorer, while the gap in income between the richest 10 % and the poorest 10 % has increased – the rich now have incomes 14-15 times those of the poor (the corresponding figure for the West is 4-6). The shadow economy now makes up an unprecedented 40-60 % of the total economy. A final negative factor is the difference in investment attractiveness of the various regions. The authors believe that the negative socio-economic trends in 1991-1999 did not result from market reforms, but rather from the lack of consistency in their implementation. The Ukraine still remains in a state of transition.

Despite some similarities between the Ukraine's regional economies (the 25 oblasts), examination reveals considerable differences in the dynamics of production, structural transformation, etc. The regions have different ratings according to development indicators. A better understanding of the essence of these processes, their determinative basis, and establishing an information base of the regional development of industry and policy planning, is only possible when the typology of old industrialised regions has been settled. Such a typology is highly complex, taking into account a number of mutually dependent and generalised indicators. By standardising and integrating indicators which characterise the concentration and accumulation of industrial production we determined the following types of oblasts – over-industrialised, highly-industrialised, industrialised and low-industrialised. The first three are highlighted in the map. Sub-types were determined according to the ratio of high-tech industries and other peculiarities in the industrial structure of the oblast. Table 1 presents subtypes of each industrial type in the region.

From its beginnings in 1990 the crisis in industry acquired a systemic character, significantly damaging the picture of territorial/regional specialisation and the correlation between individual branches and industries. Even today, a deceleration in the economic collapse does not give grounds to hope for the beginning of a process of stabilisation (or indeed growth). In 1998 industrial production in the country was 49 % of the 1990 level. Rates of fall in production vary from branch to branch. The most dramatic fall occurred in light industry (production down to 25 % of 1990 level), construction materials (23 %), machine-building and metal-processing (36 %). Slightly less affected were electricity generation (63 % of 1990), glass and tableware production (54 %), and ferrous and non-ferrous metallurgy (46 %). This unprecedented collapse in production has taken place against a backdrop of hyperinflation, reaching a peak in 1993 when consumer price indices rose by a factor of 122 for food products and by a factor of 112 for non-food items, in comparison to the previous year.

Disparate falling production levels in various industries caused substantial changes in the national industry structure as a whole, with the old industrialised regions being particularly affected. Industries experiencing the greatest decline in production tend to account for a smaller share in the overall industrial structure over the period under review, and vice versa.

Table 1. Typology of Ukraine's oblasts by level of concentration and accumulation of industry, peculiarities, ratio of high-tech branches and specific branch structure

Type	Sub-type	Oblasts
Over-industrialised	With very high proportion of coal mining, electricity generation and chemical and metallurgic industry, with predominant development of middle and low-technology branches of machine-building and metal-processing etc.	Donetsk
Highly industrialised	A. With a high proportion of coal mining, electricity generation and chemical and metallurgic industry, with predominant development of high and middle-technology branches of machine-building and metal-processing, chemicals industry etc.	Luhansk, Dniepropetrovsk, Zaporizzhya
	B. With predominant development of high and middle-technology branches of machine-building, light industry, food, chemicals, wood and other industries	Kyiv, Kharkiv, Lviv
Industrialised	A. With predominant development of high- and middle-technology branches, machine-building, food, light industry and other industries	Symy, Cherkasy, Crimea
	B. With predominant development of middle and high-technology branches of machine-building, electricity generation and petrochemical industries, as well as forestry and wood industry, food, chemicals, construction materials etc.	Khmelnitskyi, Poltava, Ivano-Frankivsk, Zakarpatska, Zhytomyr, Vinnytsya
	C. With major share of harbour-related industries, with dominance of middle-technology branches of machine-building and metal-processing, electricity generation, food, construction materials etc.	Odesa, Mykolayiv
Low-industrialised	A. Represented by middle-technology branches of machine-building and metal-processing, chemicals industry and electricity generation, local raw materials branches (wood products, construction materials etc.)	Chernihyv, Volyn, Rivne, Chernivtsyi
	B. Represented mainly by middle-technology branches of machine-building and metal-processing, local raw materials production and processing (mining of non-ore materials, non-ferrous metallurgy), food industry, construction materials, heating industries (brown coal, or oil-processing)	Kyrovohrad, Kherson
	C. Mainly represented by local raw materials processing (food industry, construction materials) with dominant share of middle-technology branches e.g. machine-building and metal-processing, light industry	Ternopil'

Source: Own research

The following unexpected conclusions can be drawn from a study of the dynamics of industrial structure and its current state:

1. The period under review saw a drastic increase in the "heaviness" of industrial structure, due to a larger share of material-consuming, energy-consuming and eco-consuming heavy industrial branches – power generation, fuel industry, ferrous metallurgy and other industries.
2. The overall scope and share of high technology production (primarily machine building) has slumped. The same is true of consumer goods manufacture (primarily light industry).

If these tendencies are not reversed in the near future the Ukraine will lose its status as a highly-industrialised country, become merely industrialised or even low-industrialised. In the current general economic crisis, and with relative liberalisation of the economy and foreign trade, the abovementioned trends in structural dynamics are connected to several factors: relatively strong demand in world markets for the products of heavy industry (electricity generation, ferrous metals, chemical products etc.) with weak demand for high technology Ukrainian products, rendering the light industry non-competitive. The export of metallurgy products is often aimed at holding on to foreign markets and ensuring foreign currency revenues at any cost, rather than promoting high economic efficiency.

Although the volume of metal product exports grew in 1999 by 26.5 per cent, its monetary value dropped by 7.5 per cent. Attempts at export expansion by lowering prices only lead to antidumping actions against Ukrainian exporters, and ultimately diminish the profitability of metallurgy products. Regional industry developments are more or less contingent on the tendencies mentioned above. The extent of the drop in industrial production and the intensity of changes in the branch structure is determined by its particular features, the ratio of high-tech branches, and other factors. Our analysis demonstrates that the smallest drop in production (index of industrial production - correlation of 1998 level to 1990 level exceeds 45 %) was seen in Odesa, Mykolayiv, Zaporizzhya, Poltava, Kyiv, Vinnytsya, Ivano-Frankovsk, Khmelnytsky, Cherkasy and Rivne oblasts (Table 2).

Table 2. Classifying of Ukraine's regions by extent of drop in industrial production, 1990-1999

Drop in industrial production	Index of industrial production (1999 to 1990, in %)	Oblasts
Large	Less than 40	Volyn, Symy, Zhytomyr, Zakarpattya, Kyrovohrad, Luhansk, Lviv, Kherson, Chernihyiv
Greater than average	40-45	Crimea, Dniepropetrovsk, Donetsk, Ternopil', Kharkiv, Chernivsti
Less than average	45-60	Vinnytsya, Ivano-Frankivsk, Kyiv, Poltava, Rivne, Khmelnytskyi, Cherkasy
Small	More than 60	Zaporizzhya, Mykolayiv, Odesa

Source: Own research, State Statistics Committee (2000)

In those regions the analysed period was also marked by the most intensive changes in the structure of branches of industry (Table 3). Notable were the rapid growth in electricity generation (all five atomic power stations are situated there), oil and oil-processing industries (Poltava, Ivano-Frankivsk and Odesa oblasts), non-ferrous metallurgy (Zaporizzhya and Mykolayiv oblasts), some branches of the chemicals industry, processing of raw materials and their transit through ports Odesa region), general chemistry (Cherkasy oblast), tire production (Kyiv oblast), food industry (Khmelnytskyi, Cherkasy, Mykolayiv oblasts) and ferrous metallurgy (Zaporizzhya, Poltava oblasts). Large areas of these regions belong to the industrialised and highly industrialised type. The other regions are characterised by large or greater than average drops in industrial production and, correspondingly, by low or less than average intensities of change in industrial branch structure. Thus Table 2 and 3 show a mainly negative correlation between the rate of industrial decline and intensity of branch structure changes.

Table 3. Classifying Ukrainian regions by intensity of changes in branch structure of the industry, 1990-1999

Intensity of change	Coefficient of relative structural change[1]	Oblasts
Low	Less than 70	Luhansk, Dniepropetrovsk, Donetsk, Kyrovohrad, Symy, Vinnytsya, Zhytomyr, Zakarpatska, Ternopil', Cherkasy, Chernivtsi, Crimea
Less than average	70-80	Kharkiv, Volyn, Lviv, Chernihiv
Higher than average	80-90	Zaporizzhya, Poltava, Rivne, Khmelnitskyi, Odesa, Kherson, Kyiv
High	More than 90	Mykolayiv, Ivano-Frankivsk

Source: Own research, State Statistics Committee (2000)

In summary, a sharp fall in industrial production did not occur in regions where heavy industries were developed, such as mining, raw material processing (local and shipped), electricity generation, or where industries with a considerable export potential were located. In regions where those branches were underdeveloped or their development was hampered by a drop of production in other sectors, a sharp decrease in production was noted.

3.2 Falling Income and Rising Unemployment

The economic depression wrought terrible social consequences. After a decade of declining living standards, the average monthly income per capita was 103 UAH

[1] Value of coefficient of relative structural movements was calculated by the following formula: $I = \Sigma \left| f_1 - f_2 \right|$, where f_1 = share of a certain industry in structure of industry in base year (1990), f_2 = share of the same industry in 1999.

in 1998 (about 37 US$ at the current exchange rate) dropping further to 91 UAH (about 25 US$) in 1999.

Transformation processes over the past few years have involved the closing of mines, meaning decreasing coal production and rising industry unemployment. These processes have aggravated social issues considerably: strikes have been held and the crime rate has risen generally. The national unemployment rate was relatively low at the beginning of the 1990s, with official estimates at no more than 0.5 % in the years up to 1995). The figure began rising in 1993. An early comprehensive survey was carried out by the International Labour Office in 1995 (ILO 1995). The data given by official state statistics and ILO differ significantly, with the estimates obtained by ILO methodology exceeding national statistics by a factor of 2. This further complicates analysis, especially as national statistics do not always take into account part-time employment, unpaid and partially paid administrative leave, short-time jobs, etc. For example, although the registered unemployment rate in Ukraine remained low at around 0.5 %, almost one in every three workers in Ukrainian factories in 1995 was in hidden unemployment. According to a 1999 economic activity survey compiled using ILO methodology, 12 % of the Ukrainian workforce is currently unemployed (Table 4).

Table 4. Unemployment rate in %, using ILO methodology

	1995	1998	1999
UKRAINE	5.6	11.3	12.0
Donetsk region	4.8	9.2	10.7

Source: State Statistics Committee (2000)

The percentage of officially registered unemployed is significantly lower. For instance, on January 1, 2001, the officially registered unemployment rate in Ukraine was only 4.2 % (Website of The State Committee of Statistics in Ukraine).

The shadow economy has reached an unprecedented scale (40-60 % of the economy). The labour market is characterised by large numbers of workers facing a general contraction of industry, leading to hidden unemployment and a growth in shadow employment. The latter may bring major or additional income to the population. Sociological studies demonstrate that the informally employed make up about 90 % of employees; among the youth of school age and students the figure is about 70 %. An estimate of the number of workers employed in the shadow economy at the beginning of 1998 was 10 million. There were 7 million officially registered individual entrepreneurs 3 million unregistered. From 1995-1998 the number of people employed in the informal sector of the economy grew by 54.4 %. The largest growth occurred in Kiyv city and Kiyv oblast, Dnipropetrovsk, Kyrovohrad, Zakarpatta and Zaporizzhya oblasts (Zhalilo et al. 1999).

Shadow employment is one result of the re-distribution of the workforce following structural changes in the national economy. Because the average pay in the formal sector remains low, informal employment allows people to make enough money to survive; it also functions as a "social stabiliser". Shadow employment provides benefits by maintaining the workforce potential, slowing down general impoverishment, supporting consumer purchasing power, providing an additional supply of consumer goods and services, and developing entrepreneurial skills. However, these positive aspects do not outweigh the negative effects of a shadow economy, such as the deformation of the labour market, weakening of the economic mechanism of labour market regulation, growth of hidden employment, income stratification and a rise in social tensions. Informal unemployment also devalues skills by offering mainly low-skilled employment and is generally associated with inferior working conditions and low social and demographic indicators.

3.3 Demographic Situation

Ukraine's population shrunk in the years following independence. From nearly 52 million earlier in the decade, the population had fallen to below 50 million by 1999. Depopulation has occured in all of Ukraine's regions. The highest rates of decrease are observed in the highly urban and old industrialised regions of Donetsk, Dniepropetrovsk, Luhansk and Kharkiv oblasts. The main reason for depopulation is a falling birth rate: from 13.3 per 1,000 in 1989 the rate fell to 7.8 per 1,000 in 1999. The lowest birth rate is in eastern Ukraine, also an old industrialised and urban area.

A drastic decline in living standards, high levels of real unemployment, substandard working conditions, a housing deficit, poor public sanitation, falling nutritional levels, insufficient or expensive medical care, and environmental hazards, have all had a harmful impact on the health of the population. This is revealed by increasing mortality rates and a decline in average life expectancy. In the period 1990-1999 the mortality rate in Ukraine increased from 12.1 to 14.8 per 1,000. An important factor in the decline in health of Ukraine's people is the crisis in the health care system, arising from a severe shortage in funding. Only 3.6 % of Ukraine's GDP is allocated to health care; comparable figures for other industrialised countries are 6.2 % in Poland, 6.5 % in Portugal, 5.9 % in England, 9.0 % in Germany, and 14 % in the USA.

Another factor in the Ukraine's demographic fluctuation is the growing mobility of the population. This can be attributed to various historical, political and environmental events (such as the Chornobyl catastrophe). The collapse of the USSR, the emergence of new independent states, and subsequent ethnic conflicts and civil wars, have all led to a massive migration of about nine million people. The disparities in living standards in the newly independent states arising from the differences in the direction and pace of reform have also stimulated internal and external migration flows. In Ukraine's case the direction of these flows has changed in the last five to six years. For instance, in 1991-92 Ukraine had a positive migration balance and a growing population. From 1994, while remaining positive, migration inflows began to fall. There were no particular tendencies of settlement in old industrialised regions in comparison to other areas.

3.4 Social Problems – Youth Problems

The ongoing processes of transition have brought rapid social change to the Ukraine, and the country now faces the problems of increasing poverty, marginalisation, unemployment, and changes in societal norms and values. Although these challenge the health and development of all age groups, young people are particularly at risk, with their often unique needs and problems as they emerge from a period of rapid growth and development. They are usually dependent on adults to meet their physical and emotional needs, and given the current pressures on parents in the Ukraine, it is not surprising that these needs are not being adequately met.

The number of crimes committed by young people has increased by 12 % in the last decade. Juvenile crime now makes up 25 % of all crimes committed in the Ukraine. The highest growth in crime levels was seen in 1999, when, for example, 105 and 96 criminal gangs were uncovered in the regions Dniepropetrovsk and Donetsk respectively. The homicide rate grew by 11 % in 1999, and now amounts to 112 per 10,000 inhabitants, 27 % were carried out by gangs from Donetsk region and 16 % by gangs from Odesa region. Unemployment is an aggravating factor in criminality for many of Ukraine's regions (especially those undergoing massive restructuring), as seen by the fact that 68 % of criminals are unemployed.

The level of education in Ukraine is comparable to standards in other developed countries when examined according to a number of parameters. Education has a great potential to facilitate Ukraine's integration into the world community. Adult literacy is estimated at 99 %. According to one index of education, the different regions in the Ukraine are fairly homogenous in the levels of education they achieve (0.887) (UNDP 1999). This same index is used to characterise countries such as Israel, Portugal and Hungary. The highest levels of education are recorded in Kharkiv, Kyiv and Lviv oblasts. However, high standards of education may not be sufficient to enable the country to overcome the ongoing economic crisis. Qualified workers are being made redundant at the same rate as blue-collar workers, and there are insufficient provisions for reemployment at sufficient pay. In old industrialised regions the situation is exacerbated by the traditional orientation towards industrial exports as a way to earn foreign currency, and no funds are invested to encourage the growth of the non-productive sectors and high-tech industries.

The Ukrainian people have a low opinion of the country's current socio-economic condition. The results of the first national opinion poll on human rights and freedom in the Ukraine, reflecting the knowledge of rights and freedoms and the degree of confidence in human rights institutions, were the following: 71.6 % of respondents disagreed with the statement that human rights and freedoms are currently observed in Ukraine. The political situation in the country is highly unstable.

The continuing fall in industrial production has not led to a proportional reduction on environmental pressures. The situation of old-industrialised regions (e.g. Donetsk region) is the most difficult and hazardous. An analysis of all factors brings the conclusion that the "social attractiveness" (a combination of all indicators benefiting human life) of old industrialised regions of Ukraine is very low (primarily Donetsk, Luhansk, Odesa), while their investment attractiveness remains at a reasonably high level.

4 Strategies for Future Development of the Ukraine and its Old Industrialised Regions

4.1 Formulating Strategies for the Country's Economic Development

Under current socio-economic conditions a positive role is played by a powerful economic framework emerging in the Ukraine through the implementation of a market regional policy model on the national level. This framework will operate both as an independent political and economic category and as a part of the national strategy for the Ukraine's socio-economic development. It is underpinned by social regional growth theories. Basic production facilities, decisive for the national industrial complex, will be set up in different regions. They will guarantee increased rates of economic development for each region and for the country as a whole (National Institute for Strategic Studies 2000).

To implement this strategy it is first necessary to examine each region's specific economic features and the potential of their natural, financial and labour resources. Secondly, current possibilities and restrictions on optimal use as well as the main challenges and priorities for a region's industries and their potential for development should be identified. Special attention should be given to the strengthening of export-oriented regions. There are two requirements for the above-mentioned strategic objectives for regional policies to be achieved. The first is the need to maintain the territorial integrity of Ukraine's economic complex; the second, in accordance with the concept for sustainable development, is the necessity of maintaining the environmental and economic balance in the regions undergoing development.

A primary element in the reform of national economic relations is the area of human resources. Sluggish economic activity and lower living standards have especially affected the region's economic strength. Resulting problems are limited consumer demand (due to lower income), an increasing number of people relying on social protection, and deteriorating health indicators. Demographic factors and professional training measures are used as subsidiary points of reference in identifying a region's production specialisation and in transforming the structure and geography of the country's economic complex.

Despite all the destructive changes, industry continues to play an important role in the production complex, influencing the social and natural environment of the country and its old industrialised regions. The modernisation and rational development of industry is an important factor in stimulating the national economy in the 21st century, thereby bringing improvements in living standards and restoring the environment. The realisation of these strategic tasks will require the development of a complex nationwide industrial policy, based on a multi-level analysis of industrial development not only in regions, but also in industrial centres and nodes (key-regions), highlighting major tendencies and major directives for future development.

The following principles should be adopted in forming a strategic perspective for development (National Institute for Strategic Studies 2000):

- account taken of the evolutionary nature of industrial production; using the experience and practice of those countries which have already entered a post-industrial stage; full accordance of perspective development of the industry, its scale and specialisation with national interests of the country in maintaining its security (economic, environmental, social, political);
- realising the development potential of different regions; rational use of resources; orientation towards growth and the effective functioning of industry; increased production quality; ensuring competitive production;
- using industrial development to strengthen the Ukraine's external role; growth of export potential of the country and its regions; rational participation in international division of labour;
- active environmental policy in industrial production; reduction of environmental pressures; strengthening of social orientation of industrial policy, bringing improvements in living standards and quality of life for the population.

Today the main task for industrial development is to lead the country out of the ongoing crisis. Problems are not only associated with the transformation of economic conditions (transition from a command to a market economy, collapse of inter-regional economic ties), but also arise from the inherent structure of the economic system (management, bureaucracy, technological, functional).

4.2 Perspectives for the Development of Old Industrialised Regions

An analysis of the peculiarities and development tendencies of industry in Ukrainian regions, when combined with the experience of other countries, allows some predictions to be made regarding the future direction of industrial development. This is of long-term importance for all regions, where development must be implemented consistently and step-by-step. At the first stage, when local and regional financial resources are mobilised, with some foreign investment and state subsidies, the following two types of production should be modernised:

1. Traditional and new export-oriented industries, ensuring an inflow of foreign currency.
2. Production of consumer goods with high liquidity and quick return of capital (food industry, some branches of light industry, construction materials production, wood processing etc.). The development of these industries will limit the import of consumer goods and capital outflow from regions, thereby increasing investment in the local economy.

A Development Strategy for the Donetsk Oblast

Situated in south-east Ukraine and with a population of more than five million, Donetsk oblast is one of the large old industrialised regions in Europe. The local economy is highly dependent on mining industries, ferrous metallurgy and other branches o heavy industry, leading to severe environmental pollution. Several factors led to an unfavourable starting position regarding the transformations since 1989/1990. The slow pace in the modernisation of production techniques kept productivity low. The use of mineral resources (mostly coal) over a long period of time has depleted available high quality resources, and the region must now restructure the coal industry. This is a complex and multi-faceted task; about 40 mines are planned for closure, with a loss of 67,000 jobs. A low share of high-tech industries and lack of development in science and innovation have contributed to an overall low level of new and innovative enterprise in the region. Furthermore, an emphasis on heavy industry has been at the expense of consumer goods production. The closure of mines, the run-down of coal production, and corresponding job losses in many branches of industry have caused mass unemployment and intense social problems; the region has become increasingly socially instable.

Against this background, a programme for "social and economic development in the region" was created at the end of the 1990s. The programme was devised by the Ministry of Economics with the goal of halting the decline in material production and establishing conditions for economic growth and financial stabilisation; elements of active social policy were also foreseen. Measures contained in the programme are to be implemented and financed by the regions. At the same time adequate government support is required in the form of special tax breaks, by the creation of special economic zones and by considerable economic investment. In 1998 the president signed a law on "special economic zones and special regimes of investment activity in the Donetsk region". According to that law, subjects of economic activity in towns and districts in the region receive subsidies and preferences when implementing investment projects directed at social problems. The council on special economic zones and special investment regimes is responsible for implementing the law and for selecting investment projects. It has already approved 85 projects (47 foreign investments) with allocated funds totalling 667 million US$. Within the framework of this programme 2,700 new jobs have been created and 4,100 jobs saved. The strategy of forming special economic zones and regions for priority development has proved to be very productive for the Donetsk region. However some improvements will be required in the future, such as better investigation and selection of projects in relation to strategic tasks. The transformation of economic structures and other strategies also needs to be considered.

Leonid Rudenko, Irina Gukalova and Gregory Pidgrushniy, National Academy of Science Kiew, Ukraine

At the second stage, as the country emerges from economic crisis, the focus should be on the modernisation and reconstruction of the main branches of industry, such as electricity generation, heating industry and construction materials production. At the same time this will serve as a powerful stimulus to middle-technology industries, and, ultimately, high-tech branches of machine-building.

At the third stage, as production growth steadies, investment must be aimed at fostering innovation, converting industrial production to new technologies and forms of organisation, and the creation of zones of technological development, e.g. innovation centres, techno-parks, and techno-polis structures. Other tried and tested strategies previously adopted in western and central European countries may prove suitable for the old industrialised regions in the Ukraine. Some of these have already been implemented, with varying degrees of success. For example an "Export and Integration Strategy" was adopted to encourage the growth of export-oriented industries; this provided an inflow of foreign currency to the state budget from the sale of goods on external markets as well as through foreign investments. The previously mentioned "strategy of forming regions of priority development and Special Economic Zones" also provides subsidies and tax incentives to problem regions and centres, to good effect. This strategy is currently being implemented in the old industrialised region of Donbas and some coastal regions (e.g. Odesa). A legal basis is provided by the Law on Special Economic Zones, with other laws on special economic zones adopted for some regions and cities.

An "Infrastructure Strategy", although not yet widely implemented in regional practice, looks highly promising. It foresees the creation of a European network of transport corridors (Ministerial Conference on Crete), some of which will pass through the Ukraine (corridors 3, 5, 9, "Europe-Asia", "Baltic Sea – Black Sea" etc.). Some of these corridors have already been completed, while others are in the last stage of development. The realisation of these projects will serve not only to stimulate regional development, but will also greatly accelerate the Ukraine's integration into Europe. With the Ukraine linked to European transport and communication networks, a dynamic recreation and tourist region can be developed in the Carpathians and high/middle-technology industries activated around the axes Uzgorod-Lviv-Ternopil-Khmelnytskyi-Vinnytsya-Cherkasy-Dnipropetrivsk-Donetsk and Lviv-Rivne-Lutsk-Zhytomyr-Kyiv-Poltava-Kharkiv. Traditional industries in Donbas and Near-Dnieper are also likely to benefit.

Other important projects in support of national economic recovery are, for example, airport modernisation; certain airports are being upgraded to "international" status (Donetsk, Dniepropetrovsk, Krivyi Rig etc.). Also the Ukraine's tradition as a country with a high level of research in science and technology (rocket-building, aircraft-building, super-solid compounds, technological equipment for some branches of industry etc.) should be developed in a strategy to encourage new forms of innovative activity – technological parks, innovation centres etc. Some first steps are being taken in this direction, with, for instance, new technological parks springing up around academic institutes in Kyiv and Kharkiv. European examples should be noted here, as they provide orientation for future development in this area. Finally, the problem of protecting the environment remains acute. Here the re-profiling of some enterprises towards the manufacture of environmental protection technology would be a positive step.

As European experience has shown, strategies must have an institutional basis, with the development and adoption of laws and other legislative acts. Some steps in this direction have already been taken in the Ukraine, such as the aforementioned laws on special economic zones, the decrees from the Cabinet of Ministers on networks of international transport corridors as well as Presidential decrees on the development of mono-functional cities. An important law dealing with the Basics of the Development Stimulation in regions and depressed territories has been drafted by the Ministry of Economy and accepted for consideration by the Ukranian Parliament.

4.3 Strategies for Integration with the EU

An intensive approach to European integration and improved bilateral relations should replace the current extensive approach (where the number of events, trips and signed agreements overshadows their practical output, and political symbolism drives attention away from constructive and beneficial co-operation). Such "renewed" activity could be aimed in several directions (National Institute 1999, 384):

(1) Development of a national strategy for European integration. Such a strategy should determine the principal steps for integration and draw up a concrete and intelligent plan of action, clearly defining the competences and responsibilities of the relevant bodies. Internal Ukrainian adaptation to European realities and the transformation of integration into a powerful generator for domestic change can be seen as the main goal of this strategy, according to President Kuchma's Decree No. 615, dated June 11, 1998, and entitled "The Strategy of Ukraine's Integration into the EU".

(2) Harmonisation of the values, norms and principles of Ukraine's political and social life with the norms and principles of the EU-community. The following steps are necessary:

- democratisation of state decision-making (through the re-training of officials and the creation of an effective mass-media, with non-governmental institutions controlling the decision-making processes);
- development of mechanisms to achieve a general national consensus on the main directions of the Ukraine's European policy;
- the creation of mechanisms for civilian control over state institutions;
- implementation of a comprehensive educational campaign for primary and high-school students (assisted by the preparation of school literature) aimed at explaining the advantages of integration into EU;
- organisation of a youth exchange programme based on existing European models;
- a state policy of encouraging foreign study, with the aim of enhancing the country‘s intellectual capital;

- organisation of a coordinated long-term "EU publicity campaign". The goal, through the skilful use of the principles of mass-media persuasion, will be the orientation of society to the benefits of integration.

(3) The search for a strategic European partner. However paradoxical it may seem, the US is the Ukraine's main lobbyist in Europe, especially in EU-Atlantic structures. The "Europeanisation" of Ukraine's continental lobby could be realised through:

- an intensification of co-operation with west-European states; establishing a wide-ranging dialogue between experts and scholars, with the aim of promoting closer ties between Ukrainian and west-European political elites;
- active participation in discussions concerning NATO's new Strategic Concept and NATO's role in Europe, with emphasis placed on strengthening the European pillar of the Alliance;
- greater dialogue with the EU and the West European Union in the areas of domestic and security policy; these institutions which will determine the future of continental security policy;
- the intensification of regional co-operation and partnership (particularly regarding border regions), with the effective resolution of regional conflicts.

(4) The creation of conditions necessary to attract European investment. This will certainly demand the most serious efforts, and implies fundamental changes in Ukrainian economic life. As a first step the interstate control of investments should be examined, with the introduction of measures to prevent the corruption of officials in servicing foreign investments in Ukraine.

The Ukraine can also take part in European strategic projects such as the previously mentioned "Trans-European Network" for the development of a continental communication system. The level and intensity of Ukrainian co-operation with western European states is an important indicator of the government's intention to continue on the road of European integration; the authorities must be willing to pursue economic and political reforms and to install democratic norms and principles into Ukrainian life. Co-operation with western European states must be seen as an active tool capable to help ensure one of the most important national state interests – survival in a competitive international environment, with prosperity and security for the Ukrainian people.

4.4 Spatial Economic Structure in Compliance with Sustainable Development Principles in Ukraine

While the typology of oblasts shown above gives a general overview of the various levels of industrialisation, the basis for any policy of sustainable development in the Ukraine will be much more complex. The authors have analysed the spatial structure of the Ukraine according to various economic features, distinguishing nine types – termed functional economic territorial entities (FETEs) – differently sized and unevenly distributed over the country. Such an analysis not only gives a clear view of the patterns of productive use, but also helps to set development priorities and define packages of constructive measures aimed at maintaining sustainable socio-economic and environmental development in particular regions. The following criteria were used to define the FETEs:

1. level of diversification of production in a territory, reflecting the saturation of various types of industrial activity;
2. dominant type of production in a territory, determining the main functional orientation and production specialisation;
3. the evolutionary character of production development in a territory, particularities its genesis (primary industrialisation, re-industrialisation etc.);
4. the state of sectoral and technological structure of production in the FETE, dominant concentration of high-tech, middle-level or low-technology branches and industries (the latter mostly in depressed regions).
5. delimitation of FETEs by controlling the expansion of the division of labour and the functional relationship between production centres.

A *separate type* of FETE is represented by the territory contaminated by the Chernobyl disaster. With radiation levels of more than 5 Curie per sq. km this territory has practically no productive use and represents an alienation zone. Apart from being a focus of scientific and technical research, some work is continuing in the station itself (which may remain open for some limited time). Further development of this territory will only be possible following a radiation clean-up and removal of the “Sarcophagus”, the remains of the former power plant.

The *first type* of FETE includes multi-functional areas with urban concentrations, such as major cities, which display inter-regional functions; this type also shows a very high level of industrial concentration, a suburban type of agriculture, and priority given to post-industrial types of production. Three areas in the Ukraine show features of this type – the regions of Kyiv, Kharkiv and Lviv. They cover considerable territories and generally coincide with the limits of urban agglomerations and zones specialising in a suburban type of agriculture. These FETEs are the motors which drive sustainable development throughout the entire country. Technological progress is most active here, and a new market infrastructure is being established in these areas, forming the basis for an effective transition to a new post-industrial level of production.

The *second type* of FETE is the most common, comprising those industrial areas which were industrialised or re-industrialised on the basis of middle-level and high-tech industries (with priority to technological modernisation and partial re-specialisation of production). These areas generally cover the territory of a city or a small number of administrative districts. They are predominantly oblast centres (e.g. Vinnytsya, Cherkasy, Mykolayiv), with some towns and surrounding districts. After the liberalisation of foreign trade and the transition to a market economy, high-tech and middle-level technology branches of industry became uncompetitive and production collapsed. Despite the current gloomy socio-economic situation and economic stagnation, these FETEs will form “nuclei for growth”, driving the innovative adaptation and expansion of production, with a new consciousness of environmental and social concerns. This presupposes industrial re-specialisation to meet current market demands at home and abroad. These FETEs must evolve into regional development "nodes" for certain market sectors, ensuring the effective functioning of regional production complexes under new economic conditions.

The *third type* of FETE is the least common, comprising industrial areas with a concentration of mining and extraction industries, raw minerals processing or related branches (with priority given to the introduction of environment-friendly technologies and a lowering of technogenic pressure on the environment). Examples of this type are Kryvy Rig and Nikopol'. These areas are smaller than the previous type and are limited to territories including a number of mining and mineral processing centres; technological processes are to blame for the poor environmental record. At the same time the relatively high demand for the products of heavy industry on the world market explains the healthier economic position of these FETEs. In view of their continued importance for the country's economy, an effort should be made to improve production standards of heavy industry to meet environmental concerns.

The *fourth type* of FETE, widely spread in eastern Ukraine, comprises industrial areas with concentrated and depressed branches of the mining industry (with priority given to branch restructuring and modernisation of production and decreasing technogenic pressure on the environment). Depressed production in these areas (e.g. Makiyivka, Gorlovka) has led to major socio-economic stagnation. Industry restructuring and the socio-economic rehabilitation of these territories will require a considerable input of resources (mostly financial); a recovery plan can be implemented within the framework of special national programmes, with the focus on sustainable development.

The *fifth type* of territory comprises zones with predominantly agrarian specialisation, well-developed food industries, light industry, the manufacture of construction materials, and other branches (with priority given to the intensification of agricultural production and the modernisation of techniques for processing local raw materials). On the whole these territories are characterised by a low level of industry and infrastructure development (mostly market infrastructure). However, the enterprises found here produce goods with a high liquidity (food) and fast capital return, thus creating favourable conditions for the socio-economic development. The development of a fully fledged market infrastructure is important for these FETEs, and improvements to the investment climate, a rise in productivity and enhanced product quality are also desirable.

The *sixth type* of FETE comprises those territories with long-term forestry resources (with priorities given to the rehabilitation of the environment and a rational use of the natural resources). Areas of this type are located mostly in the north of the country and in the Carpathians. The transition to a model of sustainable development necessitates a reduction in the industrial use of forests, despite demand for lumber.

The *seventh type* of FETE designates territories of inter-regional recreational importance (with priority given to the development and modernisation of the service industry). They are determined by the location of recreational resources, and include coastal territories with beaches and a healthy climate, mountainous areas and areas endowed with spas and curative springs. This FETE type is associated with a comparatively low technogenic pressure on the environment. Adequate support and financial investment into local industries could quickly ensure a level of sustainable development in these territories. However a favourable investment and business climate must first be established. Some efforts in this direction have already been made, for example in the free economic zone Resort 'Truskavets'.

The *eighth type* of FETE – territories with extensive nature preservation activities – are basically nature preserves with limited industrial production, such as national parks, reserves, nature monuments, etc.

The *ninth type* of FETE comprises those areas where it is not possible to determine the dominant type of production and which show instead a combination of different types of production activity. Some important old industrialised cities and regions belong to this category; thus, for example, industry, agriculture and recreation are all important to the economy of the Donetsk oblast.

It should be stressed that this typology has a predominantly functional character, allowing the forecasting of perspective developments in certain regions. However the problem remains of marrying the concepts "functional economic zones" and old industrialised regions. In our opinion the apparent difficulties are illusory – these terms define different aspects of the same phenomenon. The criteria defining an "old industrialised region" are of a historical nature. They point to long-term economic expansion and activities in a region. From this perspective almost any region of Ukraine can be characterised as old industrialised; there are indeed some regions in this category where industrial production has been in existence for a very long time – in the case of Donbas, for over 200 years. Industrial production in Kharkiv, Kyiv, Odesa, Lviv, Vinnytsya also has a lengthy tradition and these areas similarly qualify as old industrialised. While in the latter examples industry is mostly concentrated around one large city with processing forming the industrial base, Donbas is a large old industrialised region which extends from Donetsk region itself to cover some neighbouring regions (Luhansk, partially Dniepropetrovsk); industry here is based on mining. Donbas can be considered an eastern counterpart to regions in western Europe such as the Ruhr valley.

References

Doroguntsov S, Gorska O (1998) Transformation economy structure: Theory and practice, in: Ekonomika Ukrainy, No 1, pp 4 -11

Gorlenko IA, Rudenko LG (1994) Problems of territory complex development, Kiev

ILO (1995) Employment prospects declining in Ukrainian industry. ILO survey documents deepening crisis, on 6th December 1995; in: http://www.ilo.org/public/english/bureau/inf/pr/28.htm

Khristov T (1996) Regional-typological analysis of the world modernisation process, in: Vestnik Moskovskogo universiteta, Ser. 5, geography, 1996, No 4, pp 14-19

Official Site of the Special Economic Zone in Donbas, in: www.sez.donetsk.ua

Official Site of The State Committee of Statistics in Ukraine, in: www.ukrstat.gov.ua

Pidgrushniy G P (2000) Features, tendencies and perspective directions of industrial development in the Ukraine and its regions. Ukrainian Geographical Journal No 2: 13-19

State Statistics Committee (2000) Statistical yearbook of Ukraine 1999, Kyiv

National Institute for Strategic Studies (2000) The monograph of the National Institute for Strategic Studies and National Institute for Ukrainian-Russian Relations (1999) Ukraine 2000 and beyond: Geopolitical priorities and scenarios of development, Kyiv, NISS

UNDP (1999) Ukraine. Human Development Report 1998, Kyiv

Zhalilo Y, Lakiza-Sachuk N, Khomra O (1999) Workplaces in "shadow": good or bad, Den newspaper, 16.12.1999

Conclusion and Perspectives: A way forward

Bernhard Müller[1], Maroš Finka[2] and Gerd Lintz[3]

[1] Leibniz Institute of Ecological and Regional Development (IOER), Dresden, and Dresden University of Technology (TU Dresden)
[2] Slovak University of Technology, Faculty of Architecture, Bratislava
[3] Leibniz Institute of Ecological and Regional Development (IOER), Dresden

1 Rise and Decline of Industry – Affected Cities and Regions in Eleven Countries under Review

Among the many challenges presented by the transformation of the post-socialist CEE countries, surely one of the most difficult is tackling the spatial consequences of the structural change associated with dramatic industrial decline. Industrial cities and regions, which for decades had been the motors of national economic development owing to their strong industrial base, are now confronted by considerable socio-economic and environmental problems in the wake of sudden obsolescence.

Against this background, the aim of this volume has been to present the results of recent research on the particular problems of cities and regions dominated by declining industries, as well as outlining the strategies that have been adopted to cope with the structural changes triggered by the fall of the socialist systems. The study of the relevant literature and the foregoing country reports (comprising Austria, Bulgaria, Czech Republic, Germany, Hungary, Latvia, Poland, Romania, Slovakia, Slovenia and Ukraine) in combination with the local and regional case studies and related discussions within the project team and the Network of Spatial Research Institutions in Central and Eastern Europe, allow some general conclusions and recommendations to be drawn, given below (cf. Lintz, Müller and Schmude 2004).

The first section in this concluding chapter will highlight the relatively complex picture of adjustment processes and underlying problems in old industrial areas, requiring a diverse range of policies to effectively address post-socialist economic development in the CEE countries. The second section points out the relevant coping strategies, even while the emerging picture is still somewhat blurred. A general trend seems to be the use of a wide variety of strategies influenced by the general approach to transition and the EU accession process. The last section will conclude with recommendations for a way forward in post-socialist economic adjustment of old industrial areas, and highlights the requirements for future research.

2 Specific Post-Socialist Problems of Restructuring

In western industrialised countries, "old" industries (e.g. textiles, coal, steel and shipbuilding) generally suffered gradual decline during the 1970s and 1980s, with Britain conspicuous by its rather more violent restructuring process. Such structural processes are therefore not unknown and comparisons can be drawn to current problems. However the past restructuring processes of western countries differ from those which have affected the Central and Eastern European (CEE) states, primarily because the latter traditionally had a stronger concentration of, and dependency on, industries facing decline. Also a much broader range of industries faced an immediate threat, including even those in the "new" technology sector. The core of the problems of industrialised cities and regions in decline stems from their historical development, although recent difficulties were catalysed by the socio-economic changes at the end of the 20th century. Formerly prestigious sectors such as microelectronics in eastern Germany (the former German Democratic Republic) or aircraft construction in the Ukraine became exposed to the full effects of global competition at that time, and found themselves under enormous pressure to adapt as quickly as possible.

The necessity for structural adjustment was therefore not only much more urgent, but adjustments also had to be implemented to cope with dramatic changes in the political-administrative system and society at large. In Central and Eastern Europe and the Baltic states, average Gross Domestic Product (GDP) slumped in 1990 by around 7 %, in 1991 by around 11 %, and in 1992 by around 3 %. The decline was not halted before the mid-nineties (EBRD 2000). However there were considerable national differences: Poland suffered no more than two years of falling GDP, contrasting with Bulgaria's nine. But even after the decline of GDP was reversed, the slump in the relative importance of industrial production continued. Industrial employment dropped further as a direct consequence of the required increase in labour productivity.

At a local and regional level, industrial and general economic development between the CEE countries has varied greatly, owing to the respective conditions prevailing in the different industrial cities and regions at the beginning of adjustment. Some industrial cities and regions have already made considerable progress in modernising and restructuring their economic and social circumstances, in response to external stimuli and newly emerging endogenous potential. A favourable combination of factors such as access to transport facilities, levels of resources and national awareness or interest shown by foreign developers, frequently produces those conditions conducive to successfully tackling structural change and ensuring sustainable economic development. Examples are capital city regions such as Warsaw, Prague and Bratislava, and notable (new) regional growth centres within the individual states, e.g. the border triangle between Slovakia, Austria and Hungary.

However, many cities and regions with a poor combination of factors, particularly in peripheral areas away from capital cities and the EU territory, have experienced little economic renewal and are, instead, confronted primarily by the costs of restructuring without much hope of achieving economic improvement. Here, determining factors such as location, inherited economic structure and environmental quality are less favourable, making it more difficult for these areas to compete successfully and overcome their structural shortcomings. These will be the problem areas of the future, as they are facing a vicious circle of permanent stagnation or terminal decline.

Such problem areas typically include single-industry regions, such as the mining regions of Lower Lusatia near the Polish border in eastern Germany and the area around the city of Most in Northern Bohemia in the Czech Republic. Other examples include the big coal and steel producing agglomerations of Upper Silesia in Poland and the Donetsk basin in Ukraine. Outside these regions, the other type of potentially long-term problem area includes many unfavourably located and lesser-known smaller cities and regions, such as the mining areas of Johanngeorgenstadt in the Ore Mountains of Germany, Simitli in south-west Bulgaria, parts of Békés county in south-east Hungary, and Latgale region in east Latvia. Relics of the socialist policy of industrialising rural and peripheral areas, these are once more suffering from a re-confirmed peripherality. While their uncompetitive economic bases are shrinking, new investors are few and far between. Most old industries were state-imposed single operations. The dominant socialist industrial conglomerates with narrow production ranges were important providers of services for their local communities, and had a fundamental influence on the future development prospects of such economically dependent cities and regions. The collapse of these "cathedrals in the desert" (Grabher 1997) thus caused an immediate crisis in these areas.

Furthermore, decades of a socialist industrial development strategy which emphasised mining and heavy industry using antiquated technology has left some regions with serious environmental pollution, compounding other economic problems (see also Tickle and Welsh 1998, Carter and Turnock 2001). This fact further undermined the potential of attracting future investment, particularly from abroad. In such circumstances outside help from the state, for example, is a sine qua non to initiate development or to successfully implement adjustment policies. Market forces will not solve the environmental problems, while indigenous capacity is insufficient. Perhaps only main urban centres with their inherent competitive strengths can find internal solutions. In addition, many regions have a weak entrepreneurial tradition and a risk-averse population outside the economic centres, as illustrated by the significantly lower number of business start-ups, for instance.

Spatial disparities between large and small areas have been considerable. As a result of the growing polarisation of growing and declining (stagnating) local and regional economies, the CEE states face a fine balancing act of consistent support for the main growth centres, hoping for subsequent trickle-down effects, while mitigating growing social and economic disparities. The latter is essential to ensure political stability.

As already implied, many problems of industrial cities and regions are interwoven and interdependent, similar to the cumulative causation model for underdeveloped regions by Myrdal (1957). There exists a vicious circle of declining industrial output and loss of image and self-esteem among the population, in turn leading to an exodus of those highly qualified and youthful workers needed for the development of new economic structures. Areas attracting only limited external corporate interest are especially afflicted. As a result, local and regional public bodies are financially overwhelmed by the maintenance costs of massive infrastructural legacies, including dealing with derelict factories and the associated clean-up costs of contaminated sites. These problems, together with the growing indirect costs of social disintegration and a lack of economic perspectives, combine to form a complex set of problems shaped by particular local and regional conditions.

3 Necessity and Possibility of Tailor-made Strategies

As discussed above, the CEE countries in general, and specifically industrial cities and regions in these countries, have been facing very difficult conditions of structural adjustment. After the first years of transition in which macroeconomic issues dominated, awareness grew of increasing spatial disparities. Given the significant differences in economic, social and political legacies (see also Dostál 1998), it is not surprising that there should also be different strategic responses. Fundamental is the very notion of what constitutes "transition" and how to deal with it. The difference in view determines the emphasis given to each of the three main political coping strategies: (1) slowing the pace of structural change to allow a more gradual adjustment, (2) cushioning the social impact of structural change and (3) creating competitive employment. Countries which favoured gradual transition and step-by-step privatisation include Romania and Ukraine. However even Poland, after choosing the rapid "shock therapy" approach to transition, hesitated when it came to privatising the mining and steel industries in Upper Silesia. It seems that in the Ukraine state, ownership and the idea of central planning still play an important role.

Although the rationale is not always clear, there are interesting country-specific differences in the ways in which policies for aiding restructuring have been aimed at problem areas. Several countries adopted an area-based approach, designating individual regions of serious industrial decline as development areas, thus making them eligible for national assistance. In Bulgaria nine such regions have been delimited as "in industrial decline" and needing adjustment, determined using five indicators of structural deficiency: the level of industrial employment, evidence of monostructure, general and industrial employment trends, and trends in industrial output. A similar approach can be found in the Czech Republic, where two main regions 'undergoing structural change' were determined, including eight districts labelled as “undergoing difficult structural change”. They are assisted by several special development programmes. In Romania 25 ‘disadvantaged areas with severe structural problems’ (mono-industrial structures and high unemployment) have been designated. The main investment incentives offered to these regions are exemptions from various taxes: customs and excise duties, as well as taxes on profit and/or value added tax (VAT) for imported raw materials and accessories. Further incentives include grants to encourage exports. In the eastern German federal state of Saxony, the State Development Plan designates “rehabilitation and development areas” which are characterised by significant environmental damage coupled with the need for economic redevelopment. In a former uranium mining area, a model project examined how these regions could be assisted effectively using a combination of spatial planning, sectoral programmes and regional management (Müller, Rathmann and Wirth 2002).

The fact that some countries did not adopt such an area-based approach does not necessarily indicate a lack of interest in industrial problem areas. In Poland, for instance, the central government concluded a contract with the economically troubled but politically influential vojvodship of Katowice in Upper Silesia, on the regional restructuring and modernisation of the economy. Although this was at a time when the Polish vojvodships did not have their own parliaments, a number of regional actors were involved (Szepański 1998). Today, such regional contracts are a common instrument of regional policy in Poland. Although there is obviously a certain awareness of the problems of the old industrial cities and regions in Poland and elsewhere, the question remains whether all cities and regions with such difficulties can make themselves heard in their bid for support.

In most of the investigated countries a variety of strategies of state assistance and local and regional self-help can be found. However, there seems to be a clear emphasis on solutions which have already been applied in Western European countries, corresponding with the general orientation towards western economic and social models. This holds true both for strategies aimed at tackling underlying structural problems as part of the “adjustment process” and, more fundamentally, for the institutional framework within which such concrete strategies can be developed and implemented effectively. Many international donors, including the United Nations, the USA and Canada have been involved in advising and funding relevant projects. Preparations for accession to the EU naturally played the most influential role, with adoption of the EU Acquis Communautaire (negotiations about 31 regulatory chapters, covering regional policy, environment and competition policy) and the availability of pre-accession assistance given by the EU (PHARE, ISPA, SAPARD). A simple example is the introduction of systems of regions corresponding to the Nomenclature des Unités Territoriale Statistiques (NUTS). They have become important categories when analysing and tackling structural problems.

Adoption of western (specifically EU) models and ideas can be found at the level of government legislation, programmes and plans concerning urban and regional development. All these prerequisites of an effective policy had to be developed from scratch in a very short time, with the result that intent and implementation did not always coincide. It is true that the desire to improve basic infrastructure such as motor and rail links or to clean up devastated environmental areas does not need a special stimulus from abroad. However the EU's influence can be detected in the emphasis placed on such ideas as the promotion of endogenous potential, including regional human resources and small and medium enterprises (SMEs) in a framework of co-operation and sustainable development.

A positive influence can also be detected in the degree of attention given to the spatial dimension of co-operation, further highlighted by the assistance programme PHARE STRUDER (1993-2000). This supported integrated regional development programmes in selected regions, particularly those affected by restructuring problems. In Poland, for instance, the PHARE STRUDER programme is considered to be a "pioneer for various policies, strategies and programmes" (Kierzkowski 2002, 13). Having traditionally used sectoral economic policy approaches during socialism, unfamiliar spatial and regional strategies were less widespread and thus challenge traditional ways of policy making. The distinct spatial focus of EU policies has encouraged a change in policy towards cross-sectional and integrated strategies. Nevertheless, policy approaches combining different types of initiatives and different government actors are still less common than in the EU member states. Single policy approaches, often within one industry sector, are more frequent. The interdependence of different location factors, for instance, is not always obvious to policy makers. Thus, although there is some growing concern shown to improve "soft" location factors such as the environment, culture and inter-actor networks, policy makers are still more comfortable with conventional measures aimed at improving "hard" location factors, such as motorways, universities and commercial zones.

Next to the designation of special "re-/development areas", Eastern Germany is also exploring other interesting strategies based on such territorial approaches. One such example is the triangle between Dessau, Bitterfeld/Wolfen and Wittenberg in Saxony-Anhalt, where lignite mining and industrial chemistry created one of the most polluted and devastated areas in Europe. In connection to the World Exhibition in Hannover 2000, this area was subject to an innovative structural programme aimed at regional revitalisation (Leimbrock and Lintz 2003). Although unemployment is still very high, the 34 projects jointly initiated and managed as co-operative initiatives in the fields of environment, culture and economy, established a sound foundation for further development. This approach is very much based on the revolutionary ideas and experience of the International Building Exhibition (IBA) Emscher Park in the western Ruhr area of Germany. Another International Building Exhibition is currently underway in Lower Lusatia, although this is partly based on strategies of regional management and regional development concepts developed "from below". Such development concepts and regional innovation and adjustment approaches have a long tradition. As evidence from Austria shows (one of the pioneers in this regard with, e.g. the Steiermark), intensive networking between the different relevant stakeholders is necessary, and this includes municipalities and regional authorities as well as the private sector, civil society and research institutions such as universities. It is not at all surprising that such initiatives are time-consuming and costly. Moreover, although they facilitate better decision making, they cannot substitute the building of "hard" infrastructure.

Apart from stretched national resources and a lack of experience and skills in operating under the entirely new post-socialist conditions, institutional shortcomings are still the main barrier to the effective implementation of structural development measures. Although the CEE countries are seeking to improve their institutional and administrative capacity in order to advance local and regional democracy, only a few accession countries can claim to have institutions at sub-national level that are capable of formulating and implementing local and regional development strategies independently. In the interest of greater policy efficacy, the regional level in particular will need to gain in prominence. Evidence of this actually happening is provided by the extensive administrative reforms in the accession countries over recent years. In Poland, the biggest accession country, decentralisation was already underway in socialist times. In 1983, central government granted local authorities decision-making responsibilities (Balchin and Sýkora 1999). Constitutional reform in 1989 brought local autonomy. This was extended to the regional level by the 1998 reform. The 49 voivodships which had acted as central state agencies were transformed into 16 more powerful entities, characterised by a dual political structure. The provincial governor represents the national government in the provinces, which now possess both regional parliaments and regional governments. The responsibility of the latter is includes creating and implementing regional development strategies. In addition to the regions, 308 counties (and 65 cities with county status) were created. After these immense changes the political-administrative system will surely be in need of some fine tuning.

There exists ample evidence of the limited efficacy of "top-down" state planning in facilitating development. More decentralised approaches seem to promise better results. In the CEE states, greater attention will in future be given to involving key (sub-national) players in the planning process, adopting a discursive approach to planning, and intensifying local and regional input. Regional management may be the way forward here, not least as a way to reflect the EU's emphasis on regionalisation. Regional development agencies have been established in the two Czech regions worst affected by the collapse of old industrial monostructures, Most in Bohemia and Ostrava in Moravia, to help them develop a new economic role. These agencies comply with EU policies and are thus eligible for EU funding. It is not unreasonable to surmise that the availability of such funds may well have provoked the creation of such agencies; normally the Czech national government seeks to retain direct control of policies in Prague. Development agencies are also being set up at an urban level, e.g. in the municipality of Zargorje ob Savie in Central Slovenia.

EU accession and the adoption of western economic models has provided CEE countries with important stimuli in tackling the problems of old industrial cities and regions. However as termination of the PHARE STRUDER Programme showed, continued EU support for the revitalisation of old industrial cities and regions after accession is far from assured. There are no tailor-made strategies in place. The challenges (and costs) of accession may also hamper the efforts of affected countries to cope with the additional challenge of redevelopment. An interesting example is given by the special economic zones introduced in Poland and the Ukraine. They are strong instruments in establishing new enterprises, as they provide intensive tax incentives and other advantages for businesses. While this strategy was adopted in Britain in the 1980s and early 1990a as an effective state sponsored means of facilitating new local business investment in old industrial areas, it had less uptake in the rest of the EU. Following accession, Poland is now obliged to reduce subsidies offered as investment incentives to businesses.

This leads to the question of whether the EU could widen assistance to old industrial cities and regions in the newly acceded CEE countries. However, traditional EU policies do not always lend themselves for a simple application to the adjustment difficulties of former socialist countries, being historically founded on western experiences of restructuring during the 1980s. It is obvious that new dedicated solutions are needed, incorporating further research on current strategies and their practicalities of implementation. Despite many good examples and approaches demonstrating the successful handling of structural change, it remains unclear whether a broader view will automatically lead to a similarly positive assessment. Problems in implementation have hindered clear identification of the specific strategies adopted.

Finally, there are doubts regarding the political will to support old industrial areas. In the face of globalisation and increased emphasis on EU competitiveness in the aftermath of the Lisbon strategy, the promotion of structurally strong city regions as regional and/or national growth centres has become dominant (Herrschel and Newman 2002). Being neither promising growth poles nor standard peripheral areas qualifying for support, old industrial areas are in some danger of neglect by falling between two policy stools.

4 For a Way Forward in Post-socialist Economic Adjustment of Old Industrial Areas

The research presented here was carried out during an interesting pre-accession period for the CEE countries, marked by lingering economic and institutional uncertainty regarding the transformation process in general and an increasingly explicit orientation towards the EU model, particularly among the accession states. It was – and still is – a difficult task for researchers and those in positions of responsibility in the CEE countries to keep pace with the permanent and fundamental changes affecting many fields of policies relevant for the redevelopment of the cities and regions. The problems to be treated resemble moving targets – indeed the policy maker is aiming at moving goal posts.

This volume has highlighted the particularly grim fate of industrial cities and regions in the CEE countries, whose economies have faced much more ruthless restructuring processes than their western counterparts a decade or so earlier. It has become clear that these new problem areas need EU and national assistance sufficiently responsive to the often diverse range of problems, reflecting the variety of inherited national and regional circumstances. The observed variation in the economic restructuring processes amongst the CEE countries is mirrored by an equally diverse range of coping strategies. Policies have sought to tackle structural change at different political-administrative levels, and some difficulties in implementation have become apparent. This invites further research in the fostering of economic regeneration, especially in old industrial areas. The preparations in the run up to EU accession (adoption of Acquis Communautaire and pre-accession aid) had a clearly positive influence on the process of redevelopment of urban and regional industrial economies through initiatives such as the PHARE programme. This, again, highlighted the need for a regional and cross-sectoral approach to problems.

Nevertheless, it is by no way certain that these problem areas will receive the most suitable support. According to Bachtler et al. (2002, 27) the current regional policy model shows "little sensitivity to the national needs and priorities" of the accession countries. EU policy even restricts the effectiveness of special economic zones, one of the few approaches which may well be appropriate for the situation in the accession countries, even if no longer adopted in Western Europe. This is especially important in the light of the planned EU regional policy reforms for the period after 2006. Since these countries will be the main recipients of EU assistance, it would seem prudent and sensible policy to ask whether existing regional policy instruments are sufficiently geared towards the particular needs of the CEE countries. Now that the enormous social and economic consequences of economic restructuring in the old industrial (urban) areas during the 1990s has faded from collective memories, there is an acute danger that similar problems and challenges currently associated with locally and regionally concentrated large-scale deindustrialisation in the CEE countries is not given the same attention and support by EU policies. The new EU focus on supporting already strong urban growth areas to further improve international competitiveness, together with ongoing politically-motivated support for "lagging" peripheral regions, holds the danger of obscuring the problems faced by the former socialist centres of economic activity. They seem to satisfy neither of the two main criteria governing current support policy: strong inherent economic capacity, or "hopeless" backwardness and decoupling from the general development process.

The following summary statements highlight the continued challenges of formulating efficient and effective policies for economically de-based industrial centres in former socialist countries. At a conference of the Network of Spatial Research Institutions in Central and Eastern Europe held in Leipzig in October 2000, the delegates adopted a declaration entitled "The Future of Industrialised Cities and Regions" (Müller and Finka 2001), based on key findings of the FOCUS research project.

In line with the general policy goal of reducing differentials in development levels between European cities and regions, as laid down in Article 158 of the Treaty of Rome, five key recommendations can be drawn from the experience of economic adjustment and sustainable development in the CEE countries:

- Greater attention needs to be given at all levels of policymaking and international co-operation to the fate of former industrial cities and regions in Central and Eastern Europe. They are experiencing difficult conditions for future development, and include many more cities than the few frequently quoted cases. In addition, a new vision for the future is required, taking into account the varying societal and economic circumstances and the underlying socialist legacies.

- The question of how to distribute the burden of revitalising these problem cases among EU member states needs re-assessing. The costs involved often exceed the financial capacities of the new member states. National and international assistance should be tailored more specifically to the needs of the old industrial economies at a sub-national level.
- Cities and regions ought to be supported in their restructuring efforts through effective decentralisation of policy-making bodies, so that local forces can be mobilised to tackle structural problems. This requires organisational support during the early phase of devolution of power and responsibility to regional and local levels. Forty years of highly centralised authority has left its mark and there is little experience of local decision-making.
- Structural problems should be increasingly addressed by policies based on inter-departmental, cross-sectoral concepts rather than single, isolated initiatives. Such policies and strategies must be transparent, realistic and easy to implement. They need the acceptance of the different stakeholders and, especially, of the population at large. Funding should be geared less towards the regional cushioning of social hardship in the aftermath of economic decline, and more towards establishing and securing viable economic alternatives using existing or new capacity, as the situation demands. This includes a much greater concern for harmonising "visionary" regional planning and "pragmatic" regional policy, especially in terms of project funding.
- Co-operation should be encouraged between old industrial cities and regions with declining economies, in order to learn from the mistakes and effective solutions of the co-operation partners. Establishing a closer collaborative network would be one important step towards better co-operation, not least for lobbying central government and/or the EU.

An important step before implementing these recommendations is to take account of existing dialogues and exchange of experiences regarding coping strategies between the old industrial cities and regions. This includes further comparative transnational research on these issues, involving collaborative networking and co-ordinated projects.

Nevertheless the task is considerable and the path towards improving conditions for the old industrial cities and regions a long one. Many affected areas have been marginalised if not stigmatised by the new post-socialist pattern of capital investment. Language barriers, different business and social practices, varying access to relevant data and differences in the political-institutional framework make such collaboration and comparisons difficult. Yet at the same time it is only from "looking over the fence" that critical appraisals of one's own policies can be objectively made and strength and influence gained from pooling forces, just as the strong urban centres are demonstrating.

All the same, it is important to be aware of the limitations of importing successful ready-made solutions from elsewhere. Local particularities do matter, and economic opportunities vary at a small scale. Perhaps of greater use is a healthy exchange of ideas, with solutions then adapted to local conditions. New ideas may also serve as a useful challenge to established practices, encouraging new thinking over time. This includes the views of EU representatives about the suitability of established EU forms of governance and policies for the specific and varying circumstances in CEE countries, especially old industrial areas which seem to have fallen off the political radar screen.

References

Bachtler J, Downes R, Mcmaster I, Raines P, Taylor S. (2002) The transfer of EU regional policy to the countries of Central and Eastern Europe: Can one size fit all?, Report to the UK Economic and Social Research Council, European Policies Research Centre, University of Strathclyde

Balchin P, Sýkora L, with Bull G (1999) Regional policy and planning in Europe, Routledge: London New York

Carter FW, Turnock D (eds) (2001) Environmental problems in Eastern Europe, Routledge: London New York

Dostál P (1998) Democratisation, economic liberalisation, and transformational slump: a cross-sectional analysis of twenty-one post-communist countries. Environment and Planning C: Government and Policy, Vol. 16: 281-306

EBRD (European Bank for Reconstruction and Development) (2000) Transition report 2000 – Employment, skills and transition, EBRD: London

Grabher G (1997) Adaptation at the cost of adaptability? Restructuring the Eastern German regional economy. In: Grabher G, Stark, D (eds) (1997) Restructuring networks in post-socialism. Legacies, linkages, and localities, Oxford University Press: Oxford, pp 107-134

Herrschel T, Newman P (2002) Governance of Europe's city regions: planning, policy and politics, Routledge: London New York

Kierzkowski T (2002) Evaluation of socio-economic programmes in Poland – Current situation and future challenges. Conference paper, Regional Studies Association, Aix en Provence 2002, www.regional-studies-assoc.ac.uk/events/Kierzkowski.pdf

Leimbrock H, Lintz G (2003) Der "Ausnahmezustand" als Entwicklungsstrategie. Große umweltorientierte Kooperationsprojekte in ostdeutschen Problemregionen. Raumforschung und Raumordnung, Vol. 61, No. 1-2: 58-67

Lintz G, Müller B, Schmude K (2004): The future of industrialised cities and regions in Central and Eastern Europe, completely revised version of a paper given at the conference "The end of socialism…? 10 years on" of the Royal Geographical Society, Post Socialist Research Group, London, 4th-5th May, 2001, submitted for publication to Geoforum

Mayr A, Grimm F-D, Buchholz HJ (eds) (2001) Transformationsforschung: Stand und Perspektiven. Beiträge einer wissenschaftlichen Tagung. Europa Regional, Special Issue of the Zeitschrift des Instituts für Länderkunde, Leipzig, Vol. 8, No. 3/4

Müller B, Finka M (2001) How to manage urban-industrial change in Central and Eastern Europe? Declaration of the Network of Spatial Research Institutions in Central and Eastern Europe on the future of industrial cities and regions with unfavourable structural conditions. EUREG – European Journal of Regional Development, No. 9: 60-61

Müller B, Rathmann J, Wirth P (2002) Sanierung und Entwicklung umweltbelasteter Räume. Modellvorhaben in einer ehemaligen Uranbergbauregion, Frankfurt am Main

Myrdal G (1957) Economic theory and underdeveloped regions, Duckworth: London

Smith A (1996) From convergence to fragmentation: uneven regional development, industrial restructuring, and the 'transition to capitalism' in Slovakia. Environment and Planning A, Vol. 28: 135-156

Szepański MS (1998) Economic restructuring and employment promotion in the Katowice Voivodship, 1989-1996. In: Keune M (ed) (1998) Regional development and employment policy. Lessons from Central and Eastern Europe, ILO-CEET, Geneva/Budapest, pp 97-125

Tickle A, Welsh I (eds) (1998) Environment and Society in Eastern Europe, Longman: Harlow

Subject Index

Contributing Authors

Braulete, Ionica
National Institute for Research and Development in Urban and Regional Planning
N. Filipescu str. 53-55, Bucharest 02961, Romania
ionica@incdurban.ro

Dillinger, Thomas, Univ. Ass. Dr.
Institut für Städtebau und Raumplanung, Technische Universität Wien
Karlsplatz 13, 1040 Wien, Österreich
tdilling@pop.tuwien.ac.at

Finka, Maroš, Prof., M. Arch., PhD.
Insititute of Spatial Planning and Urban Design, Faculty of Architecture
Slovak University of Technology
Nam. Slobody 19, 81245 Bratislava, Slovakia
finka@fa.stuba.sk

Francis, Ilgvars, Dr.
City Development Department, Riga City Council
4 Amatu Street, LV-1050 Riga, Latvia
ilgvars@atdep.rcc.lv

Grosz, András
West Hungarian Research Institute, Centre for Regional Studies, Hungarian Academy of Sciences
Liszt F. u. 10, H-9022 Györ, Hungary
grosza@rkk.hu

Gukalova, Irina V., Dr.
Institute of Geography, Ukrainian National Academy of Sciences
Volodymyrska 44, 01034 Kyiv, Ukraine
guka@inbox.ru, ira@iatp.org.ua

Janaček, Robert, M.A.
The Office of the Bratislava Self-Governing Region
Trnavska cesta 8/A, P.O. BOX 106, 820 05 Bratislava, Slovakia
rjanacek@region-bsk.sk

Jasso, Matej, M.A., PhD.
Spectra Centre of Excellence, Faculty of Architecture, Slovak University of Technology
Nam. Slobody 19, 81245 Bratislava, Slovakia
jasso@fa.stuba.sk

Krišjane, Zaiga, Dr.
Department of Human Geography, University of Latvia
Alberta iela 10, LV 1215, Riga, Latvia
zkrisjan@lanet.lv

Lintz, Gerd, Dr
Leibniz Institute of Ecological and Regional Development
Weberplatz 1, D-01217 Dresden, Germany
G.Lintz @ioer.de

Müller, Bernhard, Prof. Dr. Dr. h. c.
Leibniz Institute of Ecological and Regional Development
Weberplatz 1, D-01217 Dresden, Germany
B.Mueller@ioer.de

Nadejde, Serban Mihai, Arch.
National Institute for Research and Development in Urban and Regional Planning
N. Filipescu str. 53-55, Bucharest 02961, Romania
snadejde@incdurban.ro

Maria Novakova
National Centre for Regional Development
16-20 Alabin Str, 1000 Sofia, Bulgaria

Okraska, Elżbieta, M.Sc.Eng.
Institute of Urban Development
ul. Cieszyńska 2, 30-015 Kraków, Poland
Z6okraska@cyf-kr.edu.pl

Pantea, Dorottya, Arch.
National Institute for Research and Development in Urban and Regional Planning
N. Filipescu str. 53-55, Bucharest 02961, Romania
pantead@incdurban.ro

Petríková, Dagmar, Assoc. Prof., M.A., PhD.
Insititute of Spatial Planning and Urban Design, Faculty of Architecture
Slovak University of Technology,
Nam. Slobody 19, 81245 Bratislava, Slovakia
petrikova@fa.stuba.sk

Pidgrushniy, Gregory P., Dr.
Institute of Geography, Ukrainian National Academy of Sciences
Volodymyrska 44, 01034 Kyiv, Ukraine
pidgr@geogr.freenet.kiev.ua

Ploštajner, Zlata
Urban Planning Institute of the Republic of Slovenia
Trnovski pristan 2, 1000 Ljubljana, Slovenia
zlata.plostajner@urbinstitut.si

Rudenko, Leonid G., Prof. Dr.
Institute of Geography, Ukrainian National Academy of Sciences
Volodymyrska 44, 01034 Kyiv, Ukraine
rudenko@geogr.freenet.kiev.ua

Schimak, Gerhard, Hon. Prof. Dr.
Institut für Städtebau und Raumplanung, Technische Universität Wien
Karlsplatz 13, 1040 Wien, Österreich
gschimak@pop.tuwien.ac.at

Schneller, Krisztian
VÁTI Hungarian Public Company for Regional Development and Town Planning
Gellérthegy u. 30-32, H-1016 Budapest, Hungary

Schmude, Karl, Dipl.-Geoökologe
Ministry for Labour, Building and Regional Development
Mecklenburg-Vorpommern, Spatial Planning Department
Schlossstraße 6-8, 19053 Schwerin
karl.schmude@am.mv-regierung.de

Spiridonova, Julia, Assoc. Prof. Dr.
National Centre for Regional Development
16-20 Alabin Str, 1000 Sofia, Bulgaria
Sp.Julia@ncrdhp.bg; Julia@Aster.net

Stejskal, Jaromír, Ing.
Institute for Spatial
Jakubské nám. 3, 601 00 Brno, Czech Republic
stejskal@saura.cz

Szymański, Przemysław, Dr. architect
Institute of Architecture and Town Planning, Lodz Technical University
Ks.Skorupki st 6/8, 90-924 Lodz, Poland
pszymarc@p.lodz.pl

Visy, Erzsébet, Dr.
VÁTI Hungarian Public Company for Regional Development and Town Planning
Gellérthegy u. 30-32. H-1016 Budapest, Hungary
evisy@vati.hu

Wagner, Helga DI
Wirtschaftskammer Wien, Abteilung für Stadtplanung und Verkehrspolitik
A-1010 Wien, Stubenring 8 - 10
Helga.Wagner@wkw.at

Wirth, Peter, Dr.
Leibniz Institute of Ecological and Regional Development
Weberplatz 1, D-01217 Dresden, Germany
P.Wirth @ioer.de

Zala, György
VÁTI Hungarian Public Company for Regional Development and Town Planning
Gellérthegy u. 30-32, H-1016 Budapest, Hungary

MIX
Papier aus verantwortungsvollen Quellen
Paper from responsible sources
FSC® C105338

If you have any concerns about our products,
you can contact us on
ProductSafety@springernature.com

In case Publisher is established outside the EU,
the EU authorized representative is:
Springer Nature Customer Service Center GmbH
Europaplatz 3, 69115 Heidelberg, Germany

Printed by Libri Plureos GmbH
in Hamburg, Germany